Daniel Brandell, Jonas Mindemark, Guiomar Hernández
Polymer-Based Solid-State Batteries

Also of interest

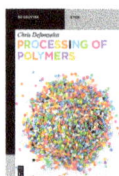

Processing of Polymers
Chris Defonseka, 2020
ISBN 978-3-11-065611-4, e-ISBN (PDF) 978-3-11-065615-2,
e-ISBN (EPUB) 978-3-11-065642-8

Electrochemistry.
A Guide for Newcomers
Helmut Baumgärtel, 2019
ISBN 978-3-11-044340-0, e-ISBN (PDF) 978-3-11-043739-3,
e-ISBN (EPUB) 978-3-11-043554-2

Electrochemical Energy Storage.
Physics and Chemistry of Batteries
Reinhart Job, 2020
ISBN 978-3-11-048437-3, e-ISBN (PDF) 978-3-11-048442-7,
e-ISBN (EPUB) 978-3-11-048454-0

Wearable Energy Storage Devices
Allibai Mohanan Vinu Mohan, 2021
ISBN 978-1-5015-2127-0, e-ISBN (PDF) 978-1-5015-2128-7,
e-ISBN (EPUB) 978-1-5015-1492-0

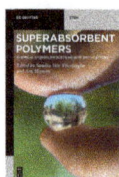

Superabsorbent Polymers.
Chemical Design, Processing and Applications
Edited by Sandra Van Vlierberghe, Arn Mignon, 2021
ISBN 978-1-5015-1910-9, e-ISBN (PDF) 978-1-5015-1911-6,
e-ISBN (EPUB) 978-1-5015-1171-4

Daniel Brandell, Jonas Mindemark,
Guiomar Hernández

Polymer-Based Solid-State Batteries

—

DE GRUYTER

Authors
Prof. Daniel Brandell
Department of Chemistry – Ångström Laboratory
Uppsala University
Box 538
SE-751 21 Uppsala
Sweden
Daniel.Brandell@kemi.uu.se

Dr. Jonas Mindemark
Department of Chemistry – Ångström Laboratory
Uppsala University
Box 538
SE-751 21 Uppsala
Sweden
jonas.mindemark@kemi.uu.se

Dr. Guiomar Hernández
Department of Chemistry – Ångström Laboratory
Uppsala University
Box 538
SE-751 21 Uppsala
Sweden
guiomar.hernandez@kemi.uu.se

ISBN 978-1-5015-2113-3
e-ISBN (PDF) 978-1-5015-2114-0
e-ISBN (EPUB) 978-1-5015-1490-6

Library of Congress Control Number: 2021936238

Bibliographic information published by the Deutsche Nationalbibliothek
The Deutsche Nationalbibliothek lists this publication in the Deutsche Nationalbibliografie;
detailed bibliographic data are available on the Internet at http://dnb.dnb.de.

© 2021 Walter de Gruyter GmbH, Berlin/Boston
Cover image: Muhammad Abdelhamid.
Typesetting: Integra Software Services Pvt. Ltd.
Printing and binding: CPI books GmbH, Leck

www.degruyter.com

Preface

The ambition of this book is to give a brief introduction to the rapidly growing field of solid-state batteries, where the liquid components in conventional lithium-ion batteries are replaced with polymeric, and thereby solid-state, materials. Solvent-free *polymer electrolytes* are thus at the focal point, and it is discussed to what extent these can substitute the otherwise widely used liquid and gel systems. In comparison to much of the classical literature on polymer electrolytes, this book tries to keep a strong focus on battery applications. In this sense, not only the structure and dynamics of the salt–polymer interactions are of interest but also the interaction with electrodes and other battery components. While polymer electrolyte is a research area that stretches back almost 50 years, it is primarily the last decade that has seen an explosion in polymer electrolyte-based battery devices – both in scientific literature and as commercial applications in industry. Also in comparison to classical literature in the field, we here try to broaden the perspectives and include a wider polymer host platform than the standard poly(ethylene oxide), which has been the main target of scientific development ever since the inception of the research field. As we will argue, stretching beyond the polyether paradigm will be necessary for future advancement in the area and for better functional solid-state batteries. The batteries targeted are primarily high-energy-density devices, resulting in a natural focus on Li-based chemistries.

We would like to thank a few people that in different ways have contributed to the accomplishment of this book: Muhammad Abdelhamid, Tim Nordh, Alexis Rucci, Mark Rosenwinkel, Monika Schönhoff and Michel Armand. Matthew Lacey and Tim Bowden coauthored a review article in *Progress in Polymer Science* with some of us a couple of years ago, which sparked our ambitions to move forward toward this book. We would also like to thank all people at the Ångström Advanced Battery Centre of Uppsala University, and especially members of the Polymer Used in Batteries group – both past and present. It is indeed a very rewarding community to be a part of.

Uppsala, March 2021
Daniel Brandell
Jonas Mindemark
Guiomar Hernández

https://doi.org/10.1515/9781501521140-202

Contents

1 Polymer electrolyte materials and their role in batteries

1.1 Battery growth

This current era is experiencing a tremendous growth in the interest and application of batteries. From being household items bought in supermarkets, batteries are rapidly becoming larger and larger in size, and thereby also more and more costly and complex to manage. This is connected to the world clearly entering a period of electrification. Electromobility of vehicles from scooters to electric flights requires high-performance energy storage, and intermittent energy sources from solar panels and wind parks need high-quality storage with a high energy efficiency. With a shortage in energy supply, energy storage units with poor conversion efficiency will have difficulty to compete with batteries, where the energy output is largely equivalent to the energy input. While large-scale storage in the grid and small-scale storage in Internet-of-things devices are rapidly growing in demand, today's exponential growth in the demand of batteries is primarily driven by the transport sector, and especially due to the similarly exponential growth in electric vehicles (EV). This trend is foreseen to dominate during the next decade [1].

For a high versatility of batteries, that is, to have the ability to use them in a wide range of products and applications, they need to be able to supply a vast amount of energy per gravimetric and volumetric unit. These concepts are known as specific energy (Wh kg^{-1}) or energy density (Wh L^{-1}). The same is true for the power density of batteries, equivalent to the energy delivered per unit of time and either weight or volume (W kg^{-1} or W L^{-1}). Since batteries can be connected in series or parallel in an electric circuit, it is not difficult to obtain a high energy or power storage capability irrespective of battery chemistry, but if the energy density is low it will result in very big or bulky battery packs. Therefore, it is vital to maximize the energy content per gravimetric and volumetric unit, in particular for mobile application where the penalty is strong for extra weight and volume.

The specific energy E_{sp} of a battery is determined by two factors: the specific capacity Q/m (Ah kg^{-1}) and the voltage U (V):

$$E_{sp} = \frac{U \times Q}{m} \tag{1.1}$$

In a battery, where the released energy is determined by redox reactions taking place in the battery electrodes, the voltage describes the potential difference between the battery electrodes – the driving force for the battery reaction – while the specific capacity is equivalent to how many times this electrochemical reaction can occur. One can make an analogy with driving in a nail with a hammer: the voltage

https://doi.org/10.1515/9781501521140-001

corresponds to the force of hitting the nail, while the capacity corresponds to how many times the nail is hit.

This partly explains why the Li-ion battery (LIB) technology has become dominant among secondary (rechargeable) batteries. It is relatively simple to find LIB electrode materials with a large voltage difference, while these materials can also store a lot of lithium, thereby providing high voltage – in fact, the highest theoretical voltage of all elements due to the low reduction potential of Li – at the same time as providing high capacity. Moreover, the small size and lightweight of Li^+ ion leads to high-energy-density electrode materials where it is relatively easy to find host structures where the Li^+ ions can jump in and out during battery charge and discharge. This, in turn, leads to batteries that can cycle for an extensive amount of cycles. Other commercial (lead–acid, nickel–cadmium, and nickel–metal hydride; see Fig. 1.1) and largely noncommercial (Na-ion, Mg-ion, Ca-ion, and Al-ion) battery chemistries often fail in one or several of these categories: compared to LIBs, they do not provide the same energy density or the same cycle life. LIBs therefore have, despite shortcomings in terms of cost and safety, grown to become a very useful and versatile battery type.

Wh/kg

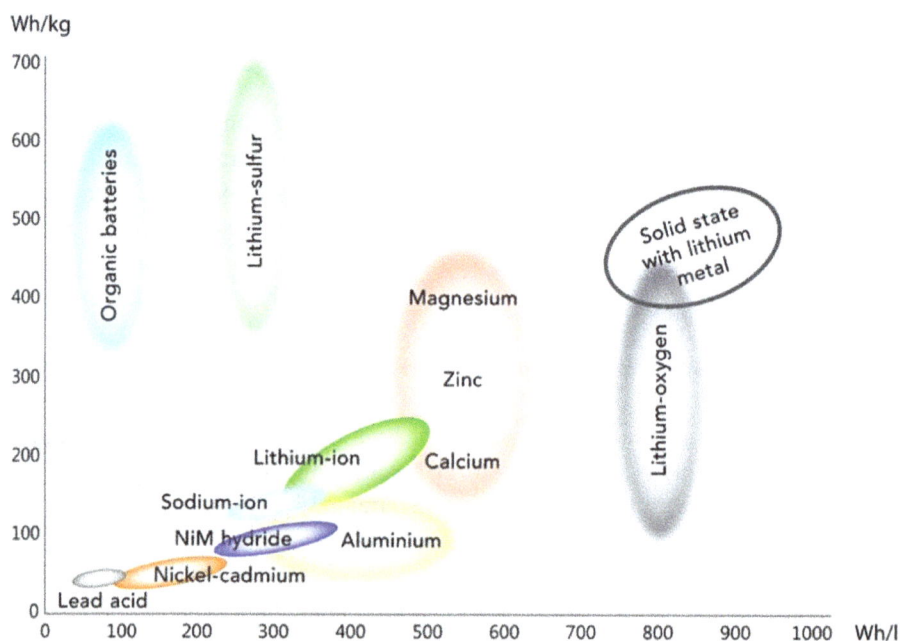

Fig. 1.1: Gravimetric and volumetric energy densities for commercial and noncommercial battery chemistries. Illustration taken from the Battery2030 + roadmap "Inventing the Sustainable Batteries of the Future. Research Needs and Future Actions."

1.2 The Li-ion battery and its electrolyte

Considering the current prominence of the LIB and its importance for the ongoing societal electrification, there will also be a natural focus on LIB chemistries throughout this book. Like all batteries, LIBs function through parallel chemical redox reactions at the two electrodes: oxidation at the anode and reduction at the cathode (Fig. 1.2). During discharge, the electrons liberated by the anode oxidation are spontaneously transported in an outer circuit over to the cathode side, where they are accepted in the reduction reaction. The electronic current this gives rise to can then be used to produce electric work. In an LIB, the anode is often graphite with Li^+ ions intercalated between the graphene sheets. During the oxidation reaction, Li^+ ions leave the electrode and travel into the electrolyte, and graphite thereby gives up one electron:

$$LiC_6 \rightarrow Li^+ + C_6 + e^-$$

The cathode, in turn, normally consists of a transition metal oxide. The transition metal ions are reduced when Li^+ ions are inserted – *intercalated* – into the host structure from the electrolyte. One common example, and the predominant cathode material in cell phone and laptop batteries, is $LiCoO_2$ (LCO):

$$CoO_2 + Li^+ + e^- \rightarrow LiCoO_2$$

There is a range of other cathode materials employed in commercial LIBs, for example, $LiFePO_4$ (LFP) and $LiNi_xMn_yCo_zO_2$ (NMC). LFP is by comparison often considered more sustainable (due to that Fe is common in the Earth's crust) and is useful for high-power applications with extensive cycling but, on the other hand, has a rather low operating voltage (ca. 3.5 V vs Li^+/Li). NMC, which exists in several different compositions (i.e., the values of x, y and z in $LiNi_xMn_yCo_zO_2$ can be varied), is dominating for EVs primarily due to its high energy density.

The role of Li^+ ions in this process is thus to charge compensate in the two different redox reactions that occur spontaneously in the anode and cathode, and which take place due to the thermodynamic driving force of the system. During charging, when energy is supplied to and stored in the battery, the reverse processes occur: Li^+ is inserted into the graphite anode, and correspondingly deinserted from the cathode material, while graphite and transition metals are reduced and oxidized, respectively. This means that an effective medium needs to transport Li^+ ions between the two electrodes during battery operation. This is the role of the *electrolyte*, which is at the focal point of this book. The electrolyte consists of a salt dissolved in a solvent; for LIBs this is a lithium salt. While the electrolyte does not store any energy in itself, it plays a vital role for current transmission in the battery cells. At the same time, the electrolyte contributes additional materials, weight and volume to the system, and thereby influences energy density, cost and sustainability in a negative way. Ideally, an electrolyte should contain as little and as simple materials as possible but still provide useful ion transport properties. From a user perspective, the battery electrolyte is a necessary evil.

The final battery cells can have different forms depending on their intended use, and these also have their pros and cons. The most common forms are cylindrical cells, prismatic cells and pouch cells. All of these, however, are based on a two-dimensional design with two electrode sheets facing each other, and the electrolyte is located in between, immersed in a separator which prevents the electrodes from making contact and thereby short-circuiting. In cylindrical and prismatic cells, these sheets are rolled or wound up. In large-scale commercial applications, such as vehicle batteries, several cells are then organized into a module, and the modules in turn placed into a battery pack. The battery pack can contain hundreds of cells and is organized for efficient power transmission and cooling of the cells during operation.

As also shown in Fig. 1.2, there are a number of different electrochemical processes that are necessary to run in parallel for the battery to work. In this context, it is of importance to acknowledge that the electrodes in the battery are composites, generally with three major components: (i) the active material undergoes the electrochemical redox reactions; (ii) an electronically conductive carbon additive; and (iii) a polymeric binder that keeps the electrode structure together. The additive and binder, similarly to the electrolyte, contribute to deadweight in the battery, and are therefore generally reduced to a few percent of the electrode content. These components form a porous mixture, and the electrolyte is immersed into voids between the particles. The major processes that need to take place for the battery to operate are thus:

- the electrochemical oxidation and reduction processes, which occur in the active material particles;
- electronic transport from the redox centers and out to the battery current collector, facilitated by the carbon additive;
- solid-state transport of lithium ions from the center of the active material particles, out toward the surface and into the electrolyte, and vice versa for the reverse process;
- diffusion and migration of lithium ions from the electrode particle surfaces in the electrolyte-filled voids of the electrode, through the bulk electrolyte in the separator, and into the pores of the counter electrode.

Depending on the current range under which the battery operates, and on the materials employed, these different processes will appear as bottlenecks for fast current transmission. This will ultimately control the power (or rate) performance in the battery, which is then controlled by kinetics rather than thermodynamics. For many LIB chemistries, however, it is common that the bulk diffusivity in the electrolyte constitutes a major contributor to the internal resistance in a cell, and thereby to energy losses and limited performance. Appropriate electrolyte materials are therefore of major importance for well-functioning cells.

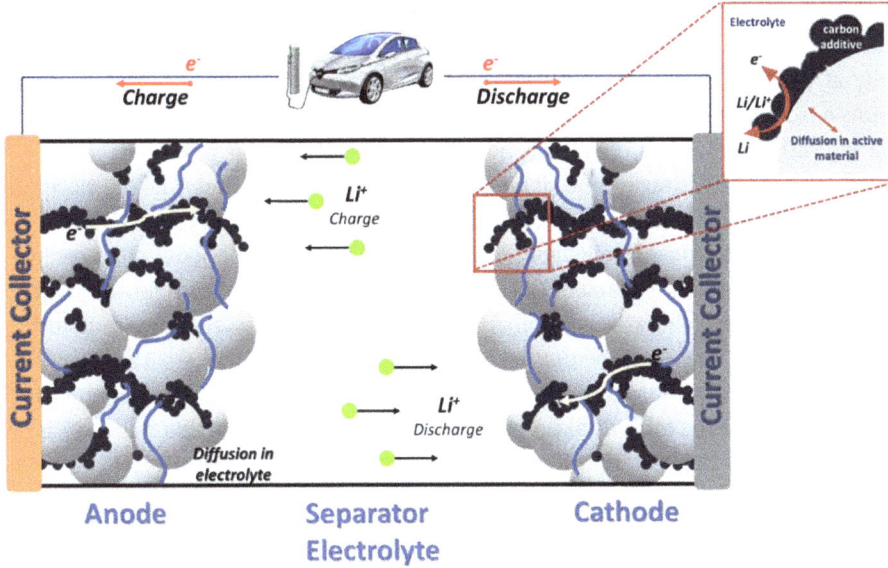

Fig. 1.2: Basic structure of the Li-ion battery cell, its components and fundamental processes. The white spheres illustrate the active anode and cathode materials, the black spheres the carbon additives used for electronic wiring, the blue strings the electrode binder and the green spheres lithium ions. The inset shows the electrochemical redox reaction and solid-state transport of lithium from the core of the active material particles.

The operating voltage U_{cell} between the two LIB electrodes is generally in the range of 3.0–4.5 V and is determined by the potential difference ΔE between the two electrodes:

$$U_{cell} = \Delta E = E_1 - E_2 \tag{1.2}$$

In comparison to other batteries – or supercaps or fuel cells, for that matter – this is a high voltage, which contributes to the high energy density of LIBs. The high and low electrode voltages correspond to the driving force of the materials to become reduced and oxidized, respectively. That this driving force is strong, analogous to that a lot of energy is being stored, also puts restrictions on the electrolyte. The electrode potentials determine the necessary *electrochemical stability window* (ESW) – that is, the potentials in between which the electrolyte does not spontaneously oxidize or reduce – which ideally should encompass the full operating voltage range of the electrodes, and thereby also exceed the battery voltage. Water – which is a common electrolyte solvent in many other battery systems – is difficult to use in LIBs, since water due to its limited ESW (1.23 V) reacts spontaneously with both the anode and cathode materials. LIBs therefore use nonaqueous electrolytes, often based on small-molecular organic solvents such as ethylene carbonate and diethylene carbonate (see Fig. 1.3) in which a lithium salt (often LiPF$_6$) is dissolved. Despite having a much wider ESW than H_2O, the electrolyte is still prone to undergoing side reactions with the LIB electrodes, primarily on the anode side. However, these side reactions normally occur only at the very beginning of battery operation, since the electrolyte decomposition forms a passivating surface layer on the surface of the anode: the *solid electrolyte interphase* (SEI). This layer is often of nanometer thickness and can be understood as an extension of the electrolyte, since it allows for ionic transport to and from the electrode particles but prevents further decomposition reactions. Despite being such a thin layer, and formed largely uncontrolledly in the LIB cell, it is a vital component that controls several key performance characteristics such as power output, battery aging and safety.

In this context, there are a number of important criteria for electrolytes to fulfill. While several of these will be discussed in greater detail in Chapters 3 and 4, a few should be highlighted already here. *Ionic conductivity*, the ability of an electrolyte to transport ions, is the most vital property of any electrolyte, since this transport is its main purpose in the battery cell. A good conductivity is directly correlated to a low internal resistance and thereby to low overpotentials and high energy efficiency. Depending on the settings of the battery management system – that is, the electronics that control the battery through different cut-off values – this will indirectly control much of the potential outtake of energy from the battery. Moreover, *stability* is a crucial parameter, since the electrolyte can decompose chemically and/or electrochemically on the electrodes. This contributes to growing internal resistance and loss of lithium, which strongly contributes to battery aging. The stability of the electrolyte decomposition products is also of major importance, as we saw above for the SEI layer. If these reaction products are well behaved in the battery cell, they can contribute to stabilizing the chemical system. *Wettability* is another crucial parameter. The electrolyte needs to fill the electrode pores and provide access for the ions to as much as possible of the surfaces of the active material particles. Without a good wettability, there will be a limited number of contact points for the electrode–electrolyte interactions, thereby generating a large interfacial resistance. Furthermore, electrolytes are vital for battery *safety*. While LIB accidents are scarce, they can be dramatic and can occur due to a number of possible failure mechanisms. In these processes, the energy stored in the electrodes is released through a cascade of reactions, where flammable material is of essence for them to proceed and accelerate. The flammable material in LIBs is essentially the organic solvents in the liquid electrolytes, which are also volatile and can cause fire to spread rapidly. Moreover, the highly fluorinated $LiPF_6$ salt and its degradation products will in this process, if exposed to water in air, decompose to HF, which is toxic [3]. Battery electrolytes therefore often contain flame retardants and other additives that mitigate different failure mechanisms. This, in turn, contributes to the *cost* of batteries, and the overall *sustainability*, which are also crucial parameters for the electrolyte from a device perspective. After the cathode material, the electrolyte is today the most costly component in an LIB [4].

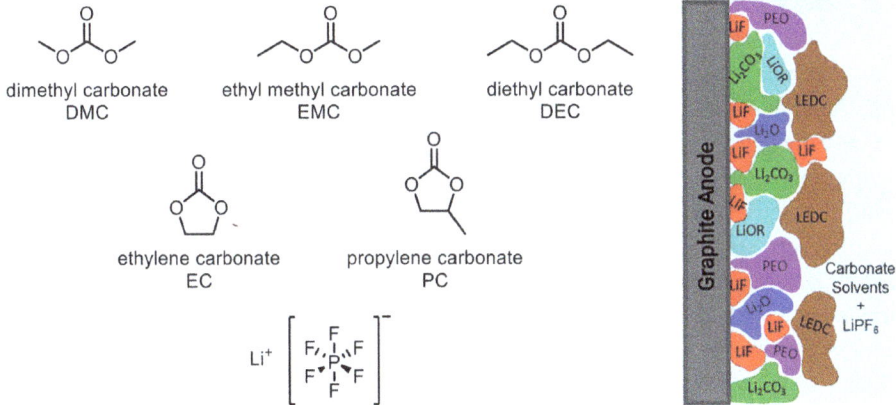

Fig. 1.3: Conventional LIB electrolyte components (left) and a simplistic picture of the SEI layer (right, reprinted from [2], Copyright 2019, with permission from Elsevier). LEDC is lithium ethylene dicarbonate and LiOR signifies lithium alkoxides.

While today's LIBs normally use the organic liquid electrolytes depicted in Fig. 1.3, it should be acknowledged that these LIBs are far from free from polymeric components. So, while LIBs can still not be considered "polymer based," they still rely heavily on polymer materials to function properly. Firstly, polymeric *binders* exist in the electrodes – for a long time considered merely as chemically inert components and not in interplay with any of the electrochemical processes in the cell – and comprise a large flora of both synthetic and natural polymers. Recent research has focused a lot on these binders, and how the binder system can be tailored for improved battery performance. Often, binders play a significant role for the surface chemistry and the electrode–electrolyte interface [5]. Secondly, the battery *separator* is generally based on polyolefins. Much work today is directed to tailor these for increased porosity, better wettability, and improved current distribution [6]. There is currently also a development of using alternative polymeric separator materials, often from biobased sources [7]. Thirdly, when the electrolyte solvents decompose to form the SEI layer, they often form polymeric components, which then cover the electrode particles. In fact, many of the additives that are put into the liquid LIB electrolytes are so-called film formers, which polymerize spontaneously at the electrolyte–electrode particle interfaces. One such important example is vinylene carbonate (VC), which forms a polycarbonate film on the anode [8] – analogous to the polycarbonate bulk electrolytes discussed later in Section 5.2 in this book. Moreover, there are plenty of polymeric materials in the packaging of the battery cells and packs.

1.3 Toward solid-state batteries

Since the non-solid component in current LIBs is the liquid electrolyte, it is the replacement of it with solid-state alternatives – ceramic or polymeric – that is necessary for creating solid-state batteries. The driving force behind this development is primarily improved safety and increased energy density, while also cost and sustainability issues, as well as a wider operational temperature range, are often addressed as potential improvements. Safety is perhaps the easiest to understand; if replacing the flammable battery component with nonflammable material, the safety hazards decrease dramatically. While many polymer materials are not exactly nonflammable, their insignificant vapor pressure makes potential risks of dramatic accidents very small. The same goes for almost all ceramic electrolyte counterparts. As the batteries are scaled up and implemented into highly safety-critical applications (e.g., marine transportation and aviation), such concerns increase in importance, and solid-state batteries are certainly targeted in this context.

The increased energy density, in turn, originates from two factors when switching to solid-state chemistries: the possibility to employ more energy-dense electrodes than in LIBs, and decreasing weight and volume of the electrolyte. The most obvious example is the employment of metal electrodes, primarily Li metal, in rechargeable batteries instead of graphite anodes. Li metal has a somewhat lower operating potential than graphite (leading to a slightly higher output voltage) but primarily a tenfold higher capacity. Li metal is, however, generally considered to be too reactive in liquid electrolytes, where the metal electrode also undergoes uneven lithium deposition during battery operation. This gives rise to needle-like structures called *dendrites* being formed, which contributes to declining battery capacity, electrolyte consumption and severe safety hazards. Solid-state electrolytes can, however, suppress dendrite formation, thereby making employment of this superior electrode possible. Moreover, also other problematic high-energy-density electrodes on both the anode and cathode sides, which normally react with liquid electrolytes, could be employed in solid-state cells: sulfur, organic electrode materials, nickel-rich cathodes, and silicon anodes [9–14]. Then, solid-state electrolytes could potentially be fabricated much thinner than conventional separators with liquid electrolytes and employ less dense materials, which would also improve energy density. But if this can be done without compromising other properties in the battery remains to be seen.

Generally, polymers are inexpensive materials that can easily be fabricated on a large scale, which likely will contribute positively to lower the material's cost in the battery. However, the largest gain in terms of both cost and sustainability is likely that many of the expensive and harmful additives in the liquid electrolytes can be omitted if the electrolyte is chemically and mechanically more robust. While this is true also for ceramic electrolytes, they generally do not possess the same low-cost potential of polymers, and most highly conductive ceramics contain some more exotic inorganic elements, for example, germanium or lanthanum, which are likely going to keep the price high. The cost benefits might therefore be less apparent for this type of solid-state chemistries. On the other hand, the wider temperature tolerance – another benefit of solid-state – is especially true for some ceramic systems, which can realize truly high-temperature batteries (>200 °C). Since LIB aging is rapidly accelerated at elevated temperatures above the preferred operating range (20–30 °C) when using liquid electrolytes, and thereby quite a lot of energy is put into battery cooling in, for example, EVs, materials that can sustain battery operating temperatures above 50 °C can actually be sought-after. Many solid-state electrolytes also do not display the same strong temperature dependence on ionic conductivity as the liquid LIB electrolytes, which also render them less temperature sensitive.

Nevertheless, despite the obvious advantages of solid-state electrolytes for batteries, they have not yet conquered much of the growing market, primarily due to two major shortcomings: ionic conductivity and electrode wettability. While wettability can be a problem for liquid electrolytes, and battery performance indeed can be improved by tailoring the surface chemistry of active materials and separators [15], these problems are much more severe for solid-state electrolytes. If porous electrodes are used, as in an LIB, the solid electrolyte first needs to fill all pores of the electrodes, which is not uncomplicated. Then, since the surface chemistry is evolving during battery operation, the electrolyte needs to be able to adapt to these changes. Most of the LIB electrode materials also change in volume during lithiation and delithiation (and conversion or alloying electrodes, e.g., Si, can experience volume changes of several hundred percent). The contraction can easily lead to loss of contact with the electrolyte if it is too rigid, while the expansion can lead to crack formation in both electrodes and electrolytes. Solid-state batteries can therefore experience very high interfacial resistances, and these problems need mitigation by thermal sintering and/or high-pressure treatments. However, since the acceptable battery lifetime is increasing to above 10 years for many applications, it is essential that the good electrode/electrolyte contacts do not degrade during battery operation. These problems are discussed more extensively in Chapter 4.

Then, as stated above, the ionic conductivity is the main property of any electrolyte system. Low ionic conductivity can be a significant problem already for liquid LIB electrolytes, where bulk transport limitations give rise to internal resistance and rate capability limitations. This is the reason for using otherwise problematic electrolyte components such as $LiPF_6$ salt: the conductivity is better as compared to most alternatives (and $LiPF_6$ also passivates the aluminum current collector well). These conductivity problems are strongly emphasized for solid-state systems, where also the conduction mechanism is fundamentally different from that in liquids. For polymer electrolytes, this is discussed in detail in Chapter 2. The result of the lower conductivity in solid-state electrolytes is higher resistances, lower energy efficiency, and generally also limited useful capacity during cycling. There is, on the other hand, an interesting interplay with the higher temperature tolerance of solid-state batteries, since ionic conductivity increases with temperature. Many solid-state electrolytes in fact display decent conductivity at elevated temperature. So, if the mechanical properties do not change dramatically and the device can operate at higher temperatures, this can be a fruitful strategy for well-performing batteries.

Yet one challenge of moving toward solid state is that the processing of batteries is required to change fundamentally. When fabricating LIBs today, the cell is usually assembled in the dry state, and the electrolyte is infiltrated right before sealing the device. While a liquid electrolyte will infiltrate the separator and electrodes fairly easily, a solid-state electrolyte will not do so within any reasonable time frame. This means that the production process for batteries needs to be rethought and reorganized. Moreover, some of these electrolyte materials are air-sensitive and/or hygroscopic, which also needs to be considered in the production process and can lead to more costly and energy-intense processing.

As already seen from the brief summary above, many of the battery properties that will change due to a transition to solid-state materials are strongly dependent on *which kind* of solid-state electrolyte materials that is envisioned. There exist solid-state electrolytes with properties that are superior to liquid electrolyte counterparts in almost every respect, but chances are small to find a single electrolyte material that surpasses the liquid LIB electrolytes in all these dimensions. While there exists a plethora of possible solid-state electrolytes, these are generally – and for good reasons – summarized into two major categories: ceramic (garnet structure oxides, phosphosulfides, perovskites, hydrides, halides, glasses, etc.) and polymeric. Sometimes, these are also referred to as "hard" and "soft" solid states, but this is somewhat of a misleading terminology since the mechanical properties vary substantially within these classes of materials, and there is a large overlap in moduli between them. Some electrolyte polymers are actually very hard while some ceramics are very soft.

Figure 1.4 summarizes qualitatively the pros and cons of the main electrolyte properties for polymeric versus ceramic electrolytes and also makes a comparison with liquid LIB counterparts. It needs to be pointed out that there are numerous exceptions to this picture, and there are several subcategories of both ceramic and polymeric conductors with fundamentally different properties than what is shown here. Nevertheless, the general picture of the ionic conductivity being the main drawback for polymer electrolytes, while interfacial compatibility being the major challenge for the ceramics, is fairly well established.

Liquid electrolyte   Polymer electrolyte   Ceramic electrolyte

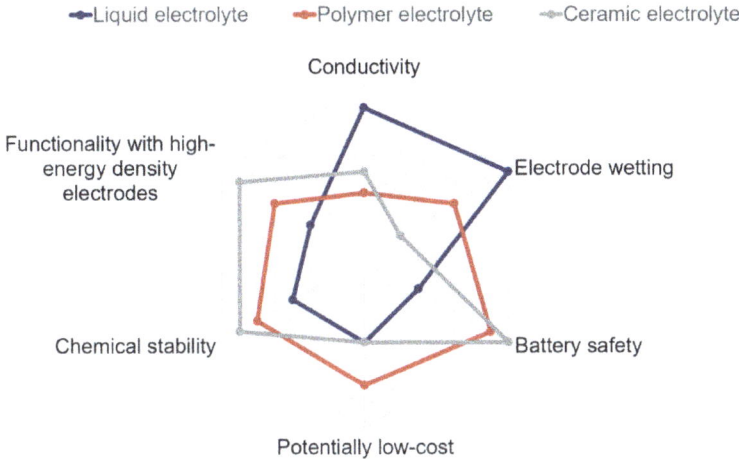

Fig. 1.4: Spider chart diagram qualitatively showing pros and cons for different electrolyte systems in batteries: conventional liquid electrolytes, solid polymer electrolytes and ceramic electrolytes.

1.4 Solid polymer electrolytes

In the following chapters, we focus on solid polymer electrolytes (SPEs), and leave ceramic electrolyte systems for others to describe in detail. We can define SPEs as *solvent-free salt solutions in polymer host materials that show sufficient mechanical stability to be considered solid in a macroscopic sense*. The scientific exploration of this type of polymer electrolytes began in the mid-1970s when P.V. Wright et al. discovered ionic conductivity in polyethylene oxide (PEO) doped with different Na and K salts [16, 17]. A few years later, M. Armand boosted the interest for these materials by targeting their use in electrochemical applications, especially Li batteries [18, 19]. Armand was also responsible for the pioneering work of attributing the ionic conductivity to the amorphous phases of the PEO:salt systems [20], which directed efforts into increasing these domains in the bulk materials by suppression of crystallinity. Previously, it had been hypothesized that the metal cations move within a helical and well-ordered PEO structure, but this picture was now replaced by envisioning the ions being transported through a constantly reforming network of coordination sites, and thereby being strongly related to the polymer segmental motion.

Relatively intense research on SPEs was carried out during the 1980s and early 1990s, and some ground-breaking efforts were made for raising the SPE conductivity, not least through the use of inorganic nanoparticles – both Li^+-conducting and insulating [21]. Polymer modifications were also looked into, using side chains and crosslinkers, in order to prevent crystallization while retaining mechanical integrity, or for inducing higher polymer flexibility. A broader range of polymer host materials than PEO was also looked into [22]. Another interesting finding at the time was that for very high salt concentrations, where the material could rather be considered a poly-mer-in-salt electrolyte (PISE) than a salt-in-polymer electrolyte, even higher conduc-tivities could be achieved. These PISE materials generally behave as plasticized salts and are often completely amorphous. Unfortunately, their mechanical behavior is less appealing, which renders them difficult to employ in most applications.

The research area had its first dedicated conference in 1987, the International Symposium on Polymer Electrolytes. Of particular interest at the time was that SPEs could be operational with Li metal in terms of sufficient electrochemical stability and prevention of excessive dendrite growth; Li metal constituted the standard Li battery anode material during these years. As stated above, this is also one of the driving forces behind the growing interest for SPEs today.

With the market introduction of LIBs based on intercalation of Li$^+$ in graphite anodes in the early 1990s [23], however, there was less interest in solving the many remaining problems for the Li metal/SPE system, and the research focus within the battery electrolyte community was instead rather drawn toward liquid systems – the currently dominating organic carbonates shown in Fig. 1.3 – where a number of differ-ent additives were implemented to stabilize the system and tailor the SEI layer forma-tion [24]. Nevertheless, SPEs continued to be an area of significant interest in academic research [25]. In the last couple of years, this area has experienced a true renaissance. With the utilization of Li and LIBs in EVs, the requirements on battery safety are stricter, and operational temperatures above ambient – contrary to the dominant usage conditions for portable electronics that constituted the bulk part of the LIB market until recently – are becoming increasingly relevant. As stated above, these requirements are appropriate for SPE materials [26]. That companies such as Bolloré have pioneered SPE battery chemistry for vehicles, thereby showing the general feasibility of solid-state ap-proaches, has certainly fueled this interest even further.

1.5 Polymer-based solid-state batteries

Li-based batteries are not the only possible application for SPEs. There exist many useful implementations for this class of materials and their gel-based counterparts in dye-sen-sitized solar cells, fuel cells, electrochromic devices, sensors, and so on, as well as for alternative battery chemistries such as Na-based systems. Nevertheless, if excluding the proton-conductive polymer membrane hosts operating with liquid components in low-temperature fuel cell electrolytes – where the ionic conductivity is strictly dependent on the presence of liquid water – the SPEs developed have most often been targeted for Li batteries. Therefore, it is primarily the conductivity of Li$^+$ ions in polymer hosts, which has attracted the most interest for these materials, and is also dominating this book.

Our focus here is on SPE-based batteries. While we build upon previous classical literature in the area that has covered different SPE materials and their properties [27, 28], the time is ripe to also incorporate these materials into a battery perspective, thereby discussing their interplay with electrode components and their behavior in battery devices. It is also due time to reflect upon the implementation of SPEs in commercial types of cells. Moreover, while previous literature has generally focused on PEO and related ether-based SPE materials, the development during the last years has rendered an increasing interest also into other polymer host materials. Therefore, we put a larger emphasis on these than what has traditionally been done. The book is organized such that we first delve into how ions are conducted in SPE materials – the very physico-chemical fundamentals behind ion transport in this category of materials. Thereafter, we discuss different techniques to analyze key electrolyte properties for SPEs while also highlighting some of the main caveats of these experimental methodologies. It is then time to discuss the behavior of SPEs in battery cells, how these devices should be understood and analyzed, and also point out a few relevant examples of such devices. A substantial part of the book is then spent on critically analyzing different types of SPE materials. These are categorized depending on their polymer host type, which is the main divider between different SPEs and what controls their ultimate performance.

We start already here in the introduction, however, by defining the borders of the SPE area somewhat strict: irrespective of their macroscopic properties, we consider SPEs being materials without any liquid components included. This restricts this category of materials to either (primarily) salts dissolved in a solid polymer host, a polymer used for plasticizing a solid salt matrix, that is, PISE materials, or polyelectrolytes containing anionic centers with coordinated metal cations. In this context, it should be acknowledged that the term "polymer electrolyte" is frequently used in the LIB field for components that are instead rather gels; that is, a liquid component or a polymer host membrane for a liquid electrolyte. Many commercial Li battery "polymer electrolytes" also contain substantial amounts of solvents or plasticizers, and the material is, again, better described as a gel (sometimes also denoted a "quasi-solid-state polymer electrolyte", if the membrane is free-standing). Irrespective of if the liquid phase is only a few percent or hundreds of percent, the membrane can macroscopically appear as a solid, but the liquid component is generally crucial for the functionality of the electrolyte. Problematically, however, it is often associated with the same stability issues and degradation processes as conventional liquid electrolytes, and many of the electrolyte properties will therefore ultimately be controlled rather by the liquid solvent component than the polymer or the salt. Safety problems are also emphasized with an increasing liquid content.

It is thus important to note the often-neglected distinction between such *gel polymer electrolytes* (GPEs) and SPEs, which is not just an issue of nomenclature, but is also significant from a more fundamental point of view in that it implies the dominant mode of ion transport: small-molecule-solvated vehicular transport in GPEs versus polymer-associated transport modes in SPEs. This is also visible from the conductivity behavior as a function of temperature (see Chapters 2 and 3). Or, more simply put: in a GPE, the ion transport is mainly related to the solvent or plasticizer rather than the polymer. This distinction also makes the ion transport in SPEs inherently more interesting from a fundamental polymer physics point of view.

This definition of SPE materials restricts what polymers can be used in solid-state batteries. They need to either be able to dissolve salts (for conventionally concentrated systems) or being able to dissolve well in salts (in highly salt-concentrated regimes), which means that the polymer needs to possess good ion-coordinating capabilities. Alternatively, for polyelectrolytes, the nonionic part of the polymer needs to have favorable interaction with the metal cations. There is, however, also a rich scientific literature on non-coordinating polymers such as poly(vinylene difluoride) (PVdF) or excessively rigid polymers such as poly(methyl methacrylate) (PMMA) used in polymer electrolyte systems but which then should have little functionality without solvents, plasticizers, and solvent residues that – perhaps unintentionally – remain after casting, or uptake of liquid from the environment during electrolyte fabrication. A high liquid content is often found also for ionomeric systems and polymerized ionic liquids, but which could – in principle – also be solid state, and are therefore discussed in Section 5.6.

Moreover, we generally leave the electrolyte materials incorporating ceramic components into the polymer matrix outside of this book. While this is a growing field in materials science [29], it is a difficult category of materials to approach, and large uncertainties exist in ion conduction mechanisms – that is, if it dominates in the polymer or ceramic phases, and how this depends on the ceramic particle loading. Results have also been conflicting in terms of how different ceramic materials interact with different polymer hosts and salt types.

This motivates our choice of materials to cover. By focusing on truly solvent-free polymeric electrolytes, comprising salts and polymers only, their behavior in batteries can be straightforwardly discussed, interpreted, and ultimately understood. This can then lay the foundation for even more complex electrolyte systems involving polymeric components.

References

[1]   World Energy Outlook 2020. International Energy Agency 2020.
[2]   Heiskanen SK, Kim J, Lucht BL. Generation and evolution of the solid electrolyte interphase of lithium-ion batteries. Joule. 2019;3:2322–33.
[3]   Larsson F, Andersson P, Blomqvist P, Mellander B-E. Toxic fluoride gas emissions from lithium-ion battery fires. Sci Rep. 2017;7:10018.
[4]   Ciez RE, Whitacre JF. Comparison between cylindrical and prismatic lithium-ion cell costs using a process based cost model. J Power Sources. 2017;340:273–81.
[5]   Bresser D, Buchholz D, Moretti A, Varzi A, Passerini S. Alternative binders for sustainable electrochemical energy storage – the transition to aqueous electrode processing and bio-derived polymers. Energy Environ Sci. 2018;11:3096–127.
[6]   Costa CM, Lee Y-H, Kim J-H, Lee S-Y, Lanceros-Méndez S. Recent advances on separator membranes for lithium-ion battery applications: From porous membranes to solid electrolytes. Energy Storage Mater. 2019;22:346–75.

[7]   Wang Z, Lee Y-H, Kim S-W, Seo J-Y, Lee S-Y, Nyholm L. Why cellulose-based electrochemical energy storage devices? Adv Mater. 2020:2000892.

[8]   Aurbach D, Gamolsky K, Markovsky B, Gofer Y, Schmidt M, Heider U. On the use of vinylene carbonate (VC) as an additive to electrolyte solutions for Li-ion batteries. Electrochim Acta. 2002;47:1423–39.

[9]   Balaish M, Peled E, Golodnitsky D, Ein-Eli Y. Liquid-free lithium–oxygen batteries. Angew Chem Int Ed. 2015;54:436–40.

[10] Villaluenga I, Wujcik KH, Tong W, Devaux D, Wong DHC, DeSimone JM, et al. Compliant glass–polymer hybrid single ion-conducting electrolytes for lithium batteries. Proc Natl Acad Sci USA. 2016;113:52.

[11] Hanyu Y, Honma I. Rechargeable quasi-solid state lithium battery with organic crystalline cathode. Sci Rep. 2012;2:453.

[12] Zheng F, Kotobuki M, Song S, Lai MO, Lu L. Review on solid electrolytes for all-solid-state lithium-ion batteries. J Power Sources. 2018;389:198–213.

[13] Manthiram A, Yu X, Wang S. Lithium battery chemistries enabled by solid-state electrolytes. Nat Rev Mater. 2017;2:16103.

[14] Gao Z, Sun H, Fu L, Ye F, Zhang Y, Luo W, et al. Promises, challenges, and recent progress of inorganic solid-state electrolytes for all-solid-state lithium batteries. Adv Mater. 2018;30:1705702.

[15] Armand M, Axmann P, Bresser D, Copley M, Edström K, Ekberg C, et al. Lithium-ion batteries – Current state of the art and anticipated developments. J Power Sources. 2020;479:228708.

[16] Fenton DE, Parker JM, Wright PV. Complexes of alkali metal ions with poly(ethylene oxide). Polymer. 1973;14:589.

[17] Wright PV. Electrical conductivity in ionic complexes of poly(ethylene oxide). Br Polym J. 1975;7:319–27.

[18] Armand M, Chabagno JM, Duclot M. Polymeric solid electrolytes. Second International Meeting on Solid Electrolytes 1978.

[19] Armand MB, Chabagno JM, Duclot MJ. Poly-ethers as solid electrolytes. In: Vashishta P, Mundy J, Gk S, eds. Fast Ion Transport in Solids: Electrodes and Electrolytes. New York, Elsevier North Holland, 1979, 131–6.

[20] Berthier C, Gorecki W, Minier M, Armand MB, Chabagno JM, Rigaud P. Microscopic investigation of ionic conductivity in alkali metal salts-poly(ethylene oxide) adducts. Solid State Ionics. 1983;11:91–5.

[21] Croce F, Appetecchi GB, Persi L, Scrosati B. Nanocomposite polymer electrolytes for lithium batteries. Nature. 1998;394:456–8.

[22] Mindemark J, Lacey MJ, Bowden T, Brandell D. Beyond PEO – alternative host materials for Li+-conducting solid polymer electrolytes. Prog Polym Sci. 2018;81:114–43.

[23] Yoshino A. The birth of the lithium-ion battery. Angew Chem Int Ed. 2012;51:5798–800.

[24] Xu K. Nonaqueous liquid electrolytes for lithium-based rechargeable batteries. Chem Rev. 2004;104:4303–418.

[25] Di Noto V, Lavina S, Giffin GA, Negro E, Scrosati B. Polymer electrolytes: Present, past and future. Electrochim Acta. 2011;57:4–13.

[26] Zhang Q, Liu K, Ding F, Liu X. Recent advances in solid polymer electrolytes for lithium batteries. Nano Res. 2017;10:4139–74.

[27] Gray FM. Polymer Electrolytes. Cambridge, RSC, 1997.

[28] Sequeira C, Santos D. Polymer Electrolytes – Fundamentals and Applications. Woodhead Publishing, 2010.

[29] Lv F, Wang Z, Shi L, Zhu J, Edström K, Mindemark J, et al. Challenges and development of composite solid-state electrolytes for high-performance lithium ion batteries. J Power Sources. 2019;441:227175.

2 Ion transport in polymer electrolytes

Strictly defined, the term *electrolyte* only refers to the dissolved, mobile ions in an electrolyte solution. This differs from the practical definition of *electrolyte* as the entire ion-conducting medium. In this text, the practical definition will generally be used unless specifically referring to an *electrolyte solution*.

2.1 Ion solvation by polymer chains

In order to generate mobile ionic species in a polymer electrolyte, the ions of a salt need to be dissolved in the polymeric solvent (commonly referred to as the *polymer host*). This is valid also for ionomers/single-ion conductors, where the counterions of the salt are tethered to the polymer chains, but where the ions are relatively immobile unless solvated to separate the oppositely charged ions. In the absence of solvation, the ion–ion interactions are much stronger than the thermal energy of the system, and there is no meaningful separation of ion pairs into free ions that can migrate in an electric field. As the cation in electrochemical systems relevant for energy storage (Li^+, Na^+, etc.) is generally much smaller and has a more localized electric charge, it has a higher tendency for strong electrostatic interactions than typical anions in Li-ion and similar battery electrolytes. The cation is therefore more critical to solvate in order to break the ion–ion interactions that stabilize the crystalline structure of the salt. This is for most SPE materials accomplished by ion–dipole interactions between the cation and Lewis basic functional groups on the polymer chains (Fig. 2.1), leading to the formation of a solvated electrolyte complex in analogy with the formation of metal complexes by the interactions of metal cations with coordinating ligands. In polymers that lack such functional groups, the necessary solvation may instead be accomplished by small-molecule additives or solvent residues. It has also been suggested that fluorophilic interactions between highly fluorinated anions and fluorinated polymer matrices also can be a driving force for salt dissolution in certain cases [1]. The structure formed by these coordinating ligands is referred to as the *solvation shell*. It is possible to define several solvation shells, referred to as the first solvation shell, second solvation shell, etc. The first solvation shell is ideally composed only of solvating ligands, resulting in full ion pair separation into "free" anions and cations, but may in practice also contain the anion directly associated to the cation, forming a *contact ion pair*. Even in a fully ion-separated system, the anion may be found not very far away in the second solvation shell, electrostatically attracted by the cation. Figure 2.2 shows the solvation shell of a Li^+ cation when coordinated by a polyester–polycarbonate copolymer host material. These cation–ligand interactions are fundamentally the same regardless of whether the solvent is a polymer or a low-molecular-weight compound. The

https://doi.org/10.1515/9781501521140-002

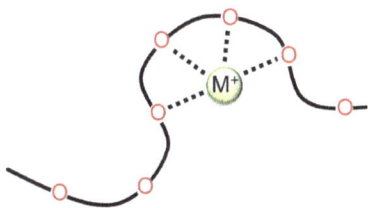

Fig. 2.1: Illustration of solvation of a metal cation (M⁺) through coordination by Lewis basic oxygen atoms in a polymer backbone.

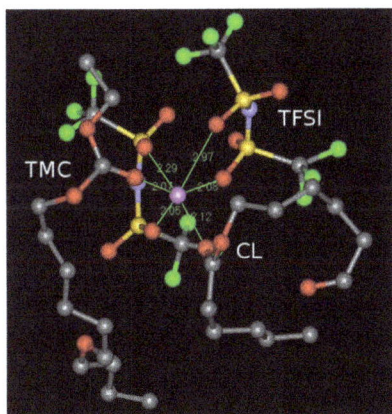

Fig. 2.2: The local coordination environment around a Li⁺ cation in a poly(ε-caprolactone-*co*-trimethylene carbonate) host matrix as obtained from molecular dynamics simulations. Adapted from [2] with permission from the PCCP Owner Societies.

polymer is, however, likely to have additional steric limitations for how the complex can be formed. Furthermore, the presence of several potentially coordinating groups on the same backbone may lead to the formation of strongly coordinating chelating structures.

In the context of ion solvation, one can make a distinction between *strong electrolytes*, which completely dissociate into free ions on dissolution, and *weak electrolytes*, which do not and where the species in solution instead comprise a significant contribution of contact ion pairs, that is, the anion is found in the first solvation shell of the cation (an example of this is shown in Fig. 2.2). It should be noted that the cation may still be solvated to some extent by the solvent as well as by the anion, and that this type of ionic aggregation should be distinguished from salt crystallization and precipitation. Depending on the degree of ionic aggregation, the solution may contain a distribution of free ions as well as ionic aggregates of various sizes and configurations, from ion pairs to triplets and larger clusters. The distinction between strongly and weakly dissociating electrolytes is dependent on the properties of the solvent as well as the salt and complete dissociation into free ions is dependent on both the solvating ability of the solvent and the strength of ion–ion interactions in the salt.

According to the hard–soft acid–base theory, small alkali metal cations are typical hard Lewis acids. This is particularly true for the Li⁺ cation, which necessitates the use of hard bases for efficient solvation. Consequently, Li⁺-conducting polymer electrolytes tend to be based on polymer hosts containing coordinating groups with highly electronegative O (ethers, carbonyls) and to a lesser degree N atoms (nitriles, amines), whereas coordination to the soft base S is not an efficient means of solvation. In line with this, many lithium salts have excellent solubility in poly(ethylene oxide), whereas poly(ethylene sulfide) is a poor solvent for the same salts [3]. The soft silver ion, on the other hand, can be favorably coordinated by S atoms, and silver complexes can, for example, be formed with poly(alkylene sulfide)s [4]. In systems with both O- and N-based solvation sites, Li^+ tends to coordinate to O, while, for example, the much softer Cu^{2+} prefers N coordination [5].

The structure of polymer hosts for cation solvation generally does not feature Lewis acidic sites that can provide strong enough interactions with anions to solvate them. In order to facilitate ion dissociation, large molecular anions with the negative charge delocalized over a large volume are therefore employed. This leads to the anions – in contrast with the solvent – being characterized by low donor numbers [6]. Typical examples include the perchlorate (ClO_4^-), tetrafluoroborate (BF_4^-), hexafluorophosphate (PF_6^-), trifluoromethanesulfonate (Tf or, more accurately, OTf⁻), bis(fluoromethanesulfonimide) (FSI) and bis(trifluoromethanesulfonimide) (TFSI) anions. The structures of the latter, more complex anions are shown in Fig. 2.3. Large atomic anions, such as the iodide anion, may also have a sufficiently dispersed negative charge to facilitate salt dissolution in polymer hosts and PEO tends to form electrolyte complexes with alkali metal iodides, whereas chlorides and bromides are less soluble [7, 8]. However, as will become clear, the SPE field today is dominated by the use of LiTFSI and LiFSI.

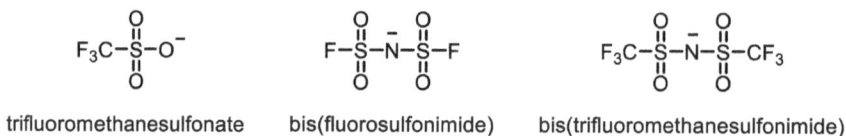

trifluoromethanesulfonate   bis(fluorosulfonimide)   bis(trifluoromethanesulfonimide)

Fig. 2.3: Structures of some common molecular anions for use in SPEs. The bis(fluorosulfonimide) and bis(trifluoromethanesulfonimide) anions are also sometimes referred to as bis(trifluorosulfon*amide*) and bis(trifluoromethanesulfon*amide*), with the abbreviations FSA and TFSA, respectively.

This dissolution of a lithium salt into a polymer host follows the same principles as the dissolution of a salt in any other solvent; the process is only favorable if there is a negative Gibbs free energy of mixing:

$$\Delta G_{mix} = \Delta H_{mix} - T\Delta S_{mix} \tag{2.1}$$

Generally, ΔS_{mix} is positive and dissolution can be thought of as an entropy-driven process – at least for mixing of low-molecular-weight components. When the solvent is polymeric, however, the situation is not as straightforward. To highlight that

ΔG_{mix} depends on both the solvent, the salt and the solution, it can alternatively be expressed as

$$\Delta G_{mix} = G_{solution} - (G_{solvent} + G_{salt}) \tag{2.2}$$

Disregarding the entropy term of Equation (2.1), this requires the enthalpy of solvation of the salt by the solvent molecules to be sufficiently large in magnitude to overcome the lattice enthalpy of the salt. In somewhat simplified terms, this essentially means that the ion–polymer interactions need to be stronger than the ion–ion (and polymer–polymer) interactions. This limits the polymer hosts to those which contain a high concentration of polar (Lewis basic) groups on the polymer chain to solvate the ions, but which at the same time are not too cohesive and rigid to allow for a reorientation and achieve a favorable ion coordination [8]. This ability is better related to the Lewis basicity of the polymer, commonly quantified as the *donor number* [9], than to the dielectric constant [10]. It is important to emphasize that the dielectric constant is a bulk property of the continuous solvent medium, and thus cannot accurately represent the molecular level, where the individual ions instead interact with local electric fields generated by dipoles in the solvent molecules. However, caution should be exercised when using donor numbers alone to assess complexation ability. As already mentioned, dissolution is dependent on a negative Gibbs free energy of the process of dissolving the salt, which necessitates to consider not only the properties of the solvent, but of the entire system both before and after ion complexation. In addition, the donor number by standard definitions does not specifically refer to coordination of Li^+, Na^+ or any other cation that is relevant for battery use (the commonly used Gutmann donor number is in fact based on solvation of $SbCl_5$ [11]) and the relative complexation strength can vary considerably between different cations.

In line with the description above, many of the polymers used as host materials for SPEs indeed have relatively low dielectric constants; PEO, for example, has a dielectric constant $\varepsilon_r \approx 5$, but is nevertheless an excellent complexing agent for Li^+. On the other hand, it could be expected that the ionic charges are poorly shielded in solutions with low dielectric constants, as the Bjerrum length (the distance at which the attraction between two oppositely charged particles is of the same magnitude as the thermal energy of the system) is inversely proportional to the dielectric constant:

$$l = \frac{e^2}{4\pi\varepsilon_0\varepsilon_r k_B T} \tag{2.3}$$

In a low-polarity solvent – such as the majority of polymer electrolyte host materials – one can thus expect a large influence of ion–ion interactions, and thereby the existence of both neutral contact ion pairs and aggregate charge carriers in the form of triplets, quintets and larger clusters. Polymer electrolytes can thus generally be thought of as weak electrolyte systems, at least at concentrations relevant for practical applications. This will, as we will see, have a profound effect on the interpretation of some of the electrochemical data obtained for SPEs.

The ion speciation, that is, the relative distribution of free ions, ion pairs, etc., is dependent on the equilibrium between fully solvated free ions and different aggregated species, which is ultimately determined by the Gibbs free energy of solvation. Strong ion–polymer interactions are equivalent to strong solvation by the host (large-magnitude free energy of solvation) and result in an equilibrium positioned toward the solvated free ions. In a system with a lower-magnitude free energy of solvation, ion–ion interactions and aggregation will instead be more favorable and the equilibrium will be shifted toward increased clustering. This equilibrium shift is exemplified in Fig. 2.4. For polymer electrolytes, salt clustering typically becomes more prominent with *increased* temperature, up until the point where salt precipitation is observed [12, 13]. This phenomenon, which often appears counter-intuitive, can be understood in the context of thermodynamics; the formation of the polymer–salt complexes leads

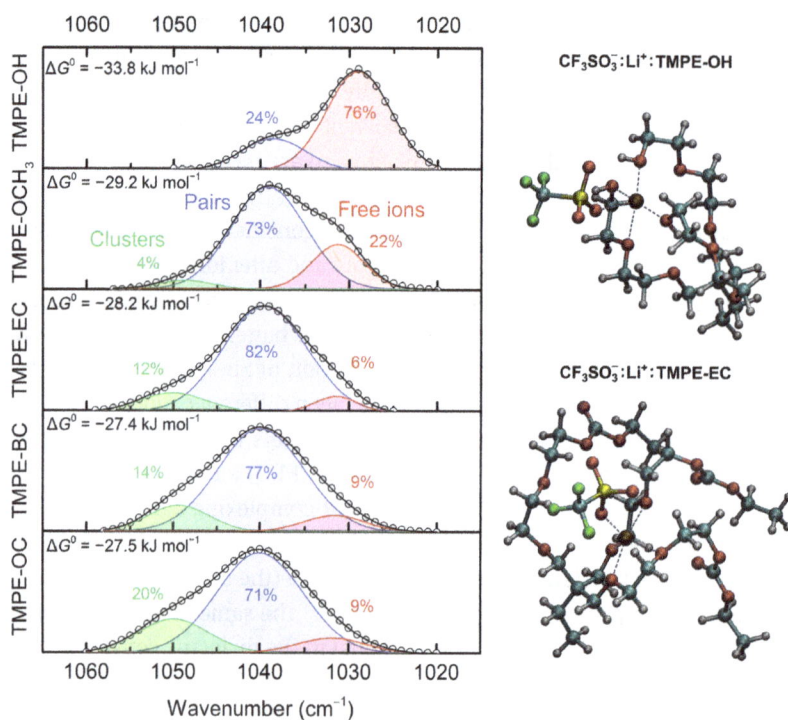

Fig. 2.4: Left: Variation of ion speciation in a series of electrolytes based on star-branched oligo (ethylene oxide) electrolytes together with LiCF$_3$SO$_3$ salt as measured by FTIR spectroscopy. The annotated percentages refer to the relative abundance of free ions, ion pairs and larger clusters. The ΔG^0 values are experimentally determined values of the Gibbs free energy of formation of the 1:1 (oligomer molecule:Li$^+$) electrolyte complexes in acetonitrile solution. -EC, -BC and -OC refer to ethyl carbonate, butyl carbonate and octyl carbonate end-groups, respectively. Right: Structures and binding motifs of two representative complexes. Adapted with permission from [19]. Copyright 2019 American Chemical Society.

to a decrease in entropy for the polymer chains and an overall unfavorable entropy of mixing. On the other hand, large ion clusters can be surprisingly stable even at high concentrations in some polymer systems, enabling the formation of polymer-in-salt electrolyte (PISE) phases, which consist, essentially, of a liquid salt phase plasticized by the polymer [14–16]. These PISE phases, also referred to as "ionic rubbers," appear at extreme salt concentrations, where the polymer in fact is the minor component, and these systems therefore behave similarly to ionic liquids [17, 18].

2.2 Fundamentals of ion transport

Ion transport in electrolyte solutions may take place through three different processes: diffusion, migration and convection. The total ion flux **J** is thus the sum of the ion flux from all these processes:

$$\mathbf{J} = \mathbf{J}_{\text{diff}} + \mathbf{J}_{\text{migr}} + \mathbf{J}_{\text{conv}} \tag{2.4}$$

In solid electrolytes, convection is not an active mode of transport – in fact, this can be considered a defining feature of a solid electrolyte – leaving diffusion and migration to carry the ion flux. While migration refers to the movement of charged species in an electric field, diffusion applies to all particles whether they are charged or neutral species and is an entropy-driven process that equilibrates differences in concentration. The diffusional flux of ions along a concentration gradient ∇c is described by Fick's first law as proportional to ∇c and a diffusion coefficient D:

$$\mathbf{J}_{\text{diff}} = -D\nabla c \tag{2.5}$$

In the absence of an electric field, ion transport in a solid electrolyte will take place solely by diffusion. When there is an electric field E (or potential gradient $\nabla\phi = -E$) present, it will exert a force F on any charged species, such as an ion, which is proportional to its charge q:

$$F = qE \tag{2.6}$$

This will cause the ion to accelerate until it reaches a steady-state drift velocity v, when a drag force that is equal in magnitude to the accelerating electrostatic force acts to limit the acceleration. This drag force is described by Stokes' law to depend on the viscosity η of the solvent and hydrodynamic radius a of the solvent–ion complex:

$$F = 6\pi\eta a \cdot v \tag{2.7}$$

From this, the drift velocity can be derived as

$$v = \frac{q}{6\pi\eta a} \cdot E \tag{2.8}$$

The initial factor in Equation (2.8) forms the electrical mobility μ and the equation can alternatively be written as

$$v = \mu \cdot E \tag{2.9}$$

While Equations (2.7) and (2.8) apply to vehicular ion transport in liquid electrolytes, the viscosity in a solid electrolyte approaches infinity, rendering Stokes' law meaningless. In addition, as we shall see later, the ion transport is not vehicular, and there is therefore little sense in considering a hydrodynamic radius. For polymer electrolytes, the dependence of ionic mobility on the dynamics of the solvent system can instead be related to the relaxation time of polymer chains.

From this line of reasoning, we can rewrite Equation (2.4), disregarding convection for solid electrolytes, as

$$J = -D\nabla c - \mu c \nabla \phi \tag{2.10}$$

Of these modes of transport, ion transport by migration is the most accessible to measure experimentally as the *ionic conductivity*. The conductivity σ of an ion i is the product of its mobility, concentration and charge, and the total ionic conductivity σ_{tot} is the sum of the contributions from all ions:

$$\sigma_{tot} = \sum \mu_i c_i |q_i| \tag{2.11}$$

However, diffusion and migration are limited by the same hydrodynamic drag (or chain mobility) and are thus interrelated, as described by the Einstein relation:

$$D = \frac{\mu k_B T}{|q|} \tag{2.12}$$

The relationship between D and μ as described by Equation (2.12) means that an electrolyte with fast ion transport by means of migration will also show (relatively) fast ion transport by means of diffusion. These processes will actually occur through the same mechanisms – it is only the driving force that differs between them. The interdependence of diffusion and migration can also be expressed through the Nernst–Einstein equation, describing the molar ionic conductivity Λ of a binary electrolyte as dependent on the diffusion coefficients of the positive and negative ions:

$$\Lambda = \frac{z^2 F^2}{RT}(D_+ + D_-) \tag{2.13}$$

Here, z represents the valence (charge number) of the ions in the salt. From Equation (2.11), it is tempting to draw the conclusion that the ionic conductivity is linearly dependent on the salt concentration, such that the conductivity can be optimized by simply saturating the salt concentration in the electrolyte solution. This is deceiving, however, as in reality the ionic mobility is not independent of salt concentration, and there is a general tendency for the mobility to decrease with higher salt concentration (more on this topic in the following section). It is also important to consider that c in Equation (2.11) refers to the concentration of charge carriers, that is, *free* ions, as opposed to neutral ion pairs. As the incidence of ion pairing becomes higher with higher salt concentration, the concentration of free ions is not a linear function of the salt concentration. Taken together, these factors lead to the presence of a conductivity maximum at some intermediate salt concentration, as illustrated in Fig. 2.5.

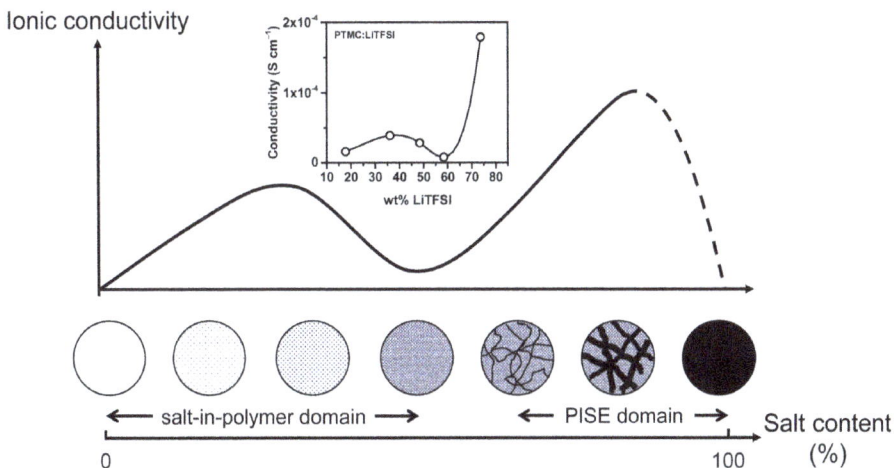

Fig. 2.5: Dependence of ionic conductivity on salt concentration for an electrolyte that transitions from a salt-in-polymer to a polymer-in-salt electrolyte as the salt clusters form a percolating network. Reprinted from [20], Copyright 2018, with permission from Elsevier. The bottom part is partially adapted from [18].

In the context of batteries, generally only the conductivity of one of the ions is of relevance for the electrochemical reactions. For applications of polymer electrolytes, this is typically the cation, for example, Li^+. To distinguish what ion species (cation or anion) is responsible for charge transport in a particular system, we may define a *transport number* t_i as the fraction of the current carried by a certain ion species i. For the cation, for example, in a Li^+-conducting system, the cation transport number t_+ is of relevance. As the transport number considers only a specific species, it is distinct from the *transference number* T_i, which, in the case of Li^+ transport, describes the number of moles of Li transferred by migration per Faraday of charge, and thus also contains contributions from ion triplets and larger ion clusters (but not ion pairs, as these are neutral and thus not transported by migration). Since negatively charged ion triplets may effectively move cations in the "wrong" direction under application of an electric field, the T_+ may indeed be negative [21], whereas t_+, which only considers the free cations, may not.

In the literature, there has long been some confusion regarding the distinction between transport and transference numbers [22]. In a sufficiently dilute electrolyte solution, where there is no ion association, $t_+ = T_+$ since all current is carried by the free ions. Since practically useful polymer electrolytes typically have much higher ion concentrations, ion association is prominent and cannot be neglected. The T_+ will thus contain contributions not only from Li^+, but also from positively and negatively charged triplets as well as larger clusters, whereas t_+ by definition only considers the free cation, making the distinction between t_+ and T_+ relevant. Despite this, these terms are commonly used interchangeably and inconsistently. Since most measurement techniques also include contributions from associated species, T_+ is probably the most relevant parameter to discuss. Alternatively, the measured values could be referred to as "apparent" or "pseudo-transference numbers" since contributions also from neutral ion pairs are often inevitably included to some extent [22].

2.3 Mechanism of ion transport in polymer electrolytes

2.3.1 Coupled ion transport

When ion conduction in polymer electrolytes was first studied in PEO-based systems, it was initially assumed that the ion transport took place by hopping between fixed coordination sites through ion channels in crystalline structures in analogy with ion transport in crystalline ceramic electrolytes [23]. On the contrary – with some exceptions (see Section 2.3.2) – it was soon discovered that the movement of ions was much more facile in – and essentially confined to – the amorphous domains of the semicrystalline polymer host [24]. A clear indicator of this is the conspicuously lower conductivity seen in semicrystalline polymer electrolytes at temperatures below the melting point, when the material crystallizes (although slow kinetics might not allow this to happen within the time frame of the measurements, as illustrated in Fig. 2.6).

Rather than consisting of ion hopping between coordination sites, ion transport in amorphous polymer electrolytes can more accurately be described as consisting of a series of ligand exchanges in a constantly evolving solvation shell of the polymer-coordinated cation (Fig. 2.7). Through this gradual evolution of the solvation shell, the cation can move between different dynamic coordination sites both along and between polymer chains. The exchange of ligands in the solvation shell of the cation thus occurs on a similar timescale as the movement of ions through the system. Since this process is directly dependent on the movements of the polymer chain itself, it is referred to as ion transport *coupled to the polymer segmental motions*. The process of dynamical rearrangements of the structure to present new favorable local environments that allow for ions to move into new coordination sites is theoretically described by the *dynamic bond percolation model*, which predicts diffusive behavior in such a system for observation times that are larger than the mean renewal time for rearrangement of the medium [25].

In the literature, it is common to find descriptions of this process as "ion hopping," but it is important to acknowledge that this mode of transport is in fact distinctly

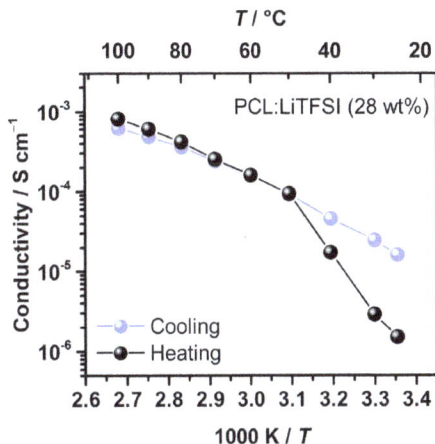

Fig. 2.6: Arrhenius plot of the total ionic conductivity of a semicrystalline electrolyte based on PCL with LiTFSI salt measured during both heating and cooling. A notable increase in conductivity can be seen as the material transforms from semicrystalline to fully amorphous at ~50 °C. During cooling, the material instead enters a metastable supercooled amorphous phase that will eventually crystallize.

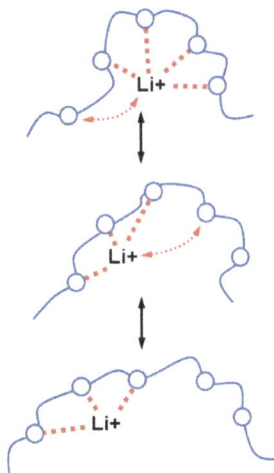

Fig. 2.7: Illustration of Li-ion movement in a generic polymer electrolyte as a series of ligand exchanges facilitated by the segmental movements of the polymer chain.

different from ion hopping between stationary coordination sites. In the latter mode of transport, the entire ion becomes essentially desolvated in the transition state between the coordination sites, whereas the cation is always in interaction with and solvated (to some extent) by the polymer in the coupled transport mode, and the transition state between coordination sites is much more loosely defined.

At the same time, ion transport coupled to the polymer segmental motions can also be sharply contrasted with the *vehicular* mode of ion transport that is active in liquid electrolytes and is characterized by ions moving together with their solvation shell; that is, the exchange of ligands in the solvation shell occurs at a timescale much slower than that of ion movement. It is obvious that this mode of transport is unreasonable in a high-molecular-weight polymer matrix, where there is virtually no diffusion of polymer chains at the relevant timescale for ion mobility. However, in oligomeric systems, vehicular transport can be noted with a gradual transition to coupled ion transport as the molecular weight increases [26]. The key phenomenon controlling this is the onset of polymer chain entanglements as the molecular weight is increased, as indicated, for example, by a sharp increase in melt viscosity (Fig. 2.8). At this point, the large-scale mobility of the polymer chains rapidly becomes extremely limited – essentially confined to reptation within a thin "tube" defined by the surrounding chains [27]. The exact point at which this occurs may be influenced by cation coordination, which changes the conformation of the chains, leading to more compact polymer coils. At sufficiently large molecular weights, the polymer chains can essentially be considered an immobile (non-diffusive) solvent on a typical experimental timescale.

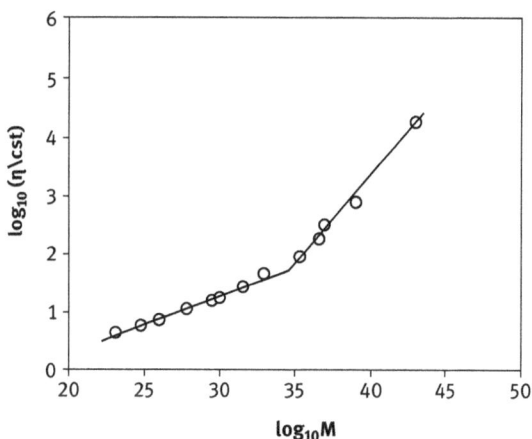

Fig. 2.8: Variation of melt viscosity with molecular weight for PEO at 100 °C, clearly showing the onset of chain entanglements. Adapted from [28], Copyright 1993, with permission from Elsevier.

This change in transport mode is reflected in the cation transport/transference number (t_+ or T_+), which decreases sharply as the coordinated cation–polymer complex becomes too large for vehicular transport, resulting in a restricted mobility of the cation relative to the anion (Fig. 2.9). In contrast to the polymer-solvated cation, the weakly coordinated anion is considered to move relatively freely with respect to the polymer chain. Some degree of coupling of the anion to the polymer is likely inevitable, however, due to a combination of electrostatic attraction between cations and anions, and the principle of excluded volume – that is, the polymer chain needs to physically move away in order to give the anion room to move within the polymer matrix.

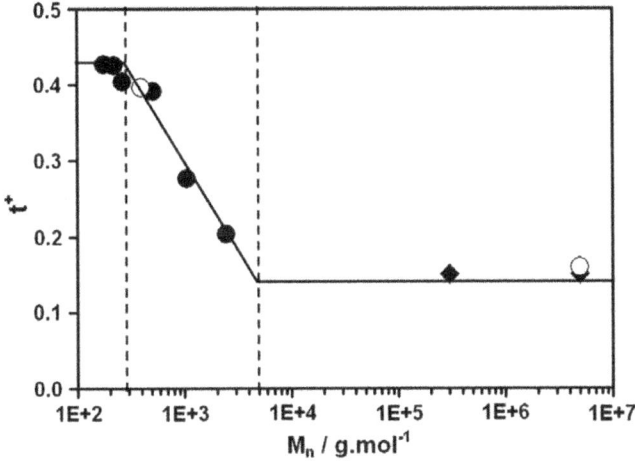

Fig. 2.9: Evolution of the cation transport number with molecular weight for poly(ethylene oxide): LiTFSI electrolytes. Reprinted from [26], Copyright 2012, with permission from Elsevier.

While the ion transport mechanisms in low-molecular-weight liquid and high-molecular-weight macromolecular electrolytes are fundamentally different, they share a similar coupling to the dynamics of the solvent molecules. As indicated by Equation (2.8), vehicular ion transport is in fact coupled to the viscosity of the electrolyte solution. This is generalized as the so-called Walden rule [29]:

$$\Lambda\eta = \text{const.} \tag{2.14}$$

In a Walden plot of molar conductivity versus viscosity, electrolytes thus typically fall on a single straight line. As already suggested, for polymer electrolytes, Equation (2.14) is rarely meaningful, and a modified Walden rule instead relates conductivity to the structural relaxation time τ_s [30]:

$$\Lambda\tau_s = \text{const.} \tag{2.15}$$

Electrolytes that deviate from this behavior can be classified as either *superionic*, with higher-than-expected conductivity, or *subionic*, when the conductivity is lower than dictated by the structural relaxations.

From the description of cation transport in polymer electrolytes as coupled to the segmental motions of the polymer host, it directly follows that ion transport can only take place in amorphous domains and only above the glass transition temperature (T_g) of the material. In this state, commonly referred to as the "rubbery state," the material may appear to be solid in a macroscopic and mechanical sense, but the local motion on a molecular level is instead essentially liquid-like. As such, the transport of ions in amorphous polymer electrolytes is closely associated with the concept of "free volume" [31] and as a general rule: the lower the glass transition temperature, the faster the segmental motions (at a given temperature) and the faster the ion transport. At T_g, the segmental mobility ceases and the ionic conductivity consequently drops sharply. While there is still some residual mobility at T_g, this ceases as the viscosity approaches infinity at the "Vogel temperature" T_0, which is experimentally found to be located ca. 50 K below T_g [31]. The connection with segmental motions and the Vogel temperature are reflected in the temperature dependence of ion conduction in polymer electrolytes not following a classic Arrhenius behavior, but instead being more accurately described by the phenomenological Vogel–Fulcher–Tammann (VFT) equation [32–34] (or, alternatively, the VFT equation, depending on how the originators are prioritized [35]):

$$\sigma = \sigma_0 \exp\left(-\frac{B}{T - T_0}\right) \tag{2.16}$$

where σ_0 and B are material-specific parameters. In this form of the equation, B has the dimension of temperature, but may alternatively be written as an energy term divided by k_B, making the equation more analogous to the Arrhenius equation. However, whereas the energy term in the Arrhenius equation is the activation energy of the process, the analogous energy term in the modified VFT equation cannot as straightforwardly be interpreted as such an activation term [8]. The pre-factor σ_0 can also be considered to be dependent on temperature, such that the equation can alternatively be written with this explicitly expressed as

$$\sigma = \frac{A}{\sqrt{T}} \exp\left(-\frac{B}{T - T_0}\right) \tag{2.17}$$

This temperature dependence is often neglected, either consciously or by ignorance. In practice, this is often found to make little difference to the fitting of experimental data and there is also some debate as to the exact temperature dependence of σ_0 [36].

The VFT model is conceptually very similar to the Williams–Landel–Ferry (WLF) model, which gives a similar temperature dependence of the ionic conductivity [37]:

$$\sigma = \sigma_0 \exp\left(\frac{-C_1(T - T_{ref})}{C_2 + T - T_{ref}}\right) \tag{2.18}$$

where T_{ref} is a reference temperature, which may be chosen to be T_g. Whereas the VFT model is derived from the variation of viscosity with temperature, the WLF model instead stems from the scaling factor that describes the temperature dependence of a segmental friction coefficient or segmental mobility [22, 37].

Importantly, the temperature dependence described by the VFT equation results in a curved line in an Arrhenius plot of log conductivity versus the inverse of temperature, whereas the straight line given by the classic Arrhenius equation is typically used to describe the conductivity in a liquid electrolyte. However, the dependence of the conductivity on viscosity in fact leads to a VFT-type conductivity behavior also in liquid systems if measured at temperatures sufficiently close to T_0.

As the temperature increases far above T_0, Equation (2.16) approaches the Arrhenius equation, as illustrated in Fig. 2.10.

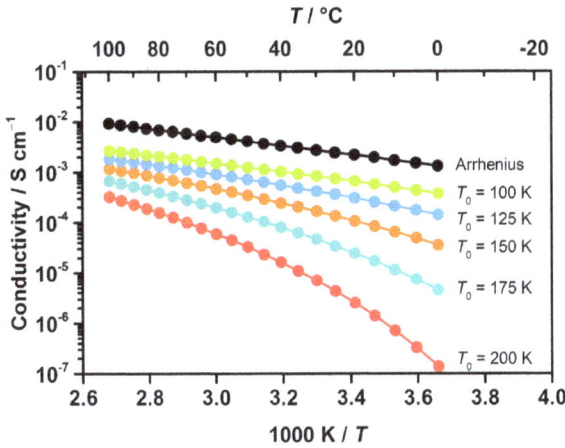

Fig. 2.10: Comparison of Arrhenius-type to VFT-type conductivity according to Equation (2.17) as T_0 falls far below T. $A = 2\ S\ cm^{-1}\ K^{1/2}$ and $B = 1{,}000\ K$. For the Arrhenius-type curve, the prefactor is $2\ S\ cm^{-1}$ and $E_A/R = 2000\ K$.

The segmental mobility of the polymer chains follows the T_g. As such, an important consideration for fast ion transport is to keep the T_g as low as possible. However, the chain mobility (and T_g) is also affected by dissolution of the salt. The solvation of cations by the polymer chains generally acts as transient physical cross-links that lower chain mobility, stiffen the material and increase the T_g. A typical example of this is Li$^+$ coordination in PEO (Fig. 2.11a). On the other hand, high concentrations of certain salts can also form large clusters that act plasticizing, thereby lowering the T_g and in turn leading to an increase in ionic conductivity. In some materials, such as poly(ethylene carbonate) (PEC) with several Li salts, this effect is seen already at fairly low salt concentrations (Fig. 2.11b) [38].

The clustering of ions to form PISEs at high concentrations may also lead to the emergence of new and efficient ion transport mechanisms in some systems. As illustrated in Fig. 2.5, the ionic conductivity initially increases with salt concentration, due to the increase in charge carrier concentration according to Equation (2.11). Coordination to the cations leads to the formation of physical cross-links that slow down the chain dynamics and causes the expected maximum in ionic conductivity to appear at a relatively moderate salt concentration. As the conductivity tapers off, however, the salt clusters formed start to dominate at higher concentrations, and may eventually reach a point where they form a percolating network [39], typically when the system contains around 50% salt. At this point, referred to as the *percolation threshold,* the properties of the system rapidly change into something resembling a plasticized salt or an ionic

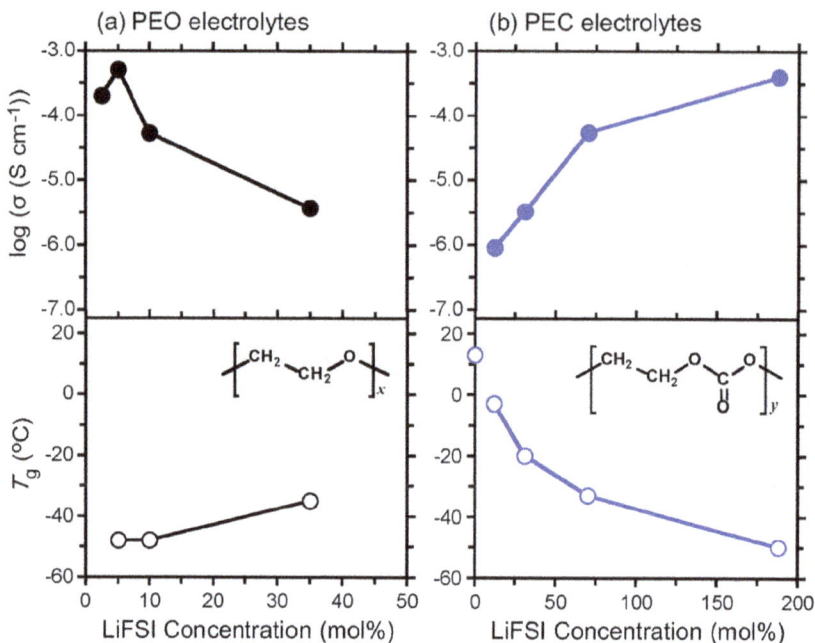

Fig. 2.11: Effects of salt concentration on T_g and ionic conductivity for (a) PEO and (b) PEC with added LiFSI salt (expressed as a molar percentage relative to the polymer host). Reprinted from [38] with permission from The Royal Society of Chemistry.

liquid [17, 18]. As a result, the ionic conductivity increases and may reach values far exceeding what can be attained in the conventional salt-in-polymer regime. While the conduction process in the PISE regime is not yet fully understood, it has been suggested to take place through hopping between salt clusters within a continuous network throughout the material [39], reminiscent of the transport of Li$^+$ in ionic liquids [40]. This is also similar to the Grotthuss mechanism that describes the rapid transport of H$^+$ and OH$^-$ in aqueous systems by short H-bond-mediated hops of protons between water molecules. The resemblance to ionic liquids unfortunately also extends to the mechanical properties of the material, which tend to deteriorate when the polymer is only present as a minor component in the system. These materials are therefore sticky, difficult to handle and often deteriorate when implemented in electrochemical devices. This naturally limits the practical utility of PISEs.

The importance of fast local chain dynamics for efficient ion transport has led to efforts to lower the T_g of the electrolyte system being a widespread strategy to prepare materials with fast ion dynamics, and thereby to SPEs with the desired high conductivity. This can be achieved by, for example, grafting ion-coordinating chains to a highly flexible (but not sufficiently ion-coordinating) backbone, such as oligo(ethylene oxide)-grafted polysiloxanes [41–43] or polyphosphazenes [44], or by incorporating flexible and plasticizing side chains that can markedly increase the overall chain dynamics [45]. However, as for PISE systems, there often exists a trade-off between the conductivity and the desired mechanical properties for a solid electrolyte.

While the segmental motions of the polymer chains are an important factor for ionic conductivity, molecular-level structural factors may also influence the rate of ion transport in the system – that is, how the ion-coordinating parts of the polymer are located with respect to each other. Considering the process of conformational rearrangements necessary to present new coordination environments, it is important that the polymer chains can easily assemble and bring the coordinating groups together in arrangements favorable for ion solvation. This can be referred to as the *connectivity of solvation sites*. It has recently been demonstrated that this is strongly dependent on the configuration and architecture of the polymer chains such that, for example, nonfunctional spacers [46] or bulky side groups [47] in the structure may impede ion transport, even if these structural elements act to increase the overall segmental mobility and lower the T_g. Such effects may thus effectively negate efforts to increase the ionic conductivity by modifying the polymer chains for fast segmental motions.

From the description of ion transport as a series of ligand exchanges, it also follows that fast cation transport is dependent on facile desolvation of the ion by individual ligands to lower the energy barrier for ligand exchange as the local coordination environment is changed. This means that the ion binding strength is an important parameter for cation transport (but not for anion movements). Accordingly, systems that are characterized by weak ion–polymer interactions show much higher relative cation mobility and thus higher cation transference numbers than polymers with a high ion binding strength [48]. The excellent solvation of Li^+ cations is thus the cause of the very low transference numbers observed for PEO and similar oxyethylene-based polyethers, whereas much higher T_+ values are reported, for example, for weakly cation-interacting polyesters and polycarbonates.

In summary, this provides a complex pattern for molecular design of the polymer host material; while it needs to have coordinating capability to dissolve the salt, the coordination strength should not be too high so that the cation is too strongly complexed. While segmental motion and low T_g are useful for ionic conductivity, high chain mobility has a tendency to decrease the mechanical properties. Moreover, tailoring the system through advanced polymer architectures can additionally disrupt the connectivity necessary for ionic migration.

2.3.2 Decoupled ion transport

The coupling of ion dynamics to the chain dynamics of the polymer host acts as a fundamental limitation to the rate of ion transport that can be achieved. To achieve ionic conductivities beyond what is dictated by the structural dynamics, *decoupled* ion transport is thus necessary. As already mentioned, ionic conductivities beyond what is dictated by Equation (2.15) are known as *superionic* and constitute an attractive proposal to realize practical utility of polymer electrolytes for applications that require much faster ion transport than what has been achieved with conventionally operating materials. Breaking the coupling between segmental dynamics and ion transport would also potentially enable fast ion transport in rigid matrices that have much more useful mechanical properties than the typical soft, amorphous polymer electrolytes.

As discussed in the previous section, the coupled ion transport mode causes the ionic conductivity to rapidly drop to a negligible level as the temperature nears the T_g. While this is the typical case, this is not really observed in certain materials, where instead measurable conductivity levels can be detected in the vicinity (or even at) T_g. This phenomenon can be quantified through the *decoupling index,* defined as

$$R_\tau = \frac{\tau_s}{\tau_\sigma} \tag{2.19}$$

The decoupling index compares the structural relaxation time τ_s with the conductivity relaxation time τ_σ at T_g. While R_τ is often below 1 for conventional SPEs, it can be considerably higher in some glassy polymer systems [49–51].

The idea of ion conduction in SPEs originally went from assuming conduction in crystalline phases to the realization that ion transport instead takes place in the amorphous regions. However, it was later demonstrated that some specific crystalline phases formed from low-molecular-weight PEO in stoichiometric complexes with alkali metal salts indeed can transport Li^+, Na^+, K^+ and Rb^+ cations [52–55]. In these structures, the PEO chains wrap in a tunnel-shaped configuration around the cations, creating structured ion transport channels with the anions located on the outside of these formations (Fig. 2.12). The cations are considered to move by hopping along the channels between coordination sites (Fig. 2.13) [56]. While it was originally surmised that ion transport in such structures would be highly selective for cation transport [52], it was later demonstrated that ion transport in the PEO_8:$NaAsF_6$ crystalline electrolyte is in fact dominated by the anions, particularly at elevated temperatures (t_+ as low as 0.17 at 40 °C) [55].

Decoupled ion transport can also be seen in some PISEs; since ion transport in these materials takes place through a continuous percolation network of ions, it is independent of the polymer host (which is perhaps more aptly referred to as a polymer *guest* in these systems), leading to values of R_τ as high as 10^{13} being noted.

In the absence of coupling of the ion transport to the segmental motions of the polymer host, there is no reason to expect VFT-type conductivity behavior, and indeed the conventional Arrhenius equation better describes the temperature dependence in rigid decoupled systems. While the ion transport mechanism in these materials has not been definitely determined, the observed behavior can be readily understood in the context of a mechanism involving ion hopping between stationary coordination sites in a rigid matrix, where the desolvated ion serves as a high-energy transition state that limits the rate of the ion transport process so that such materials show large similarities to ceramic electrolytes.

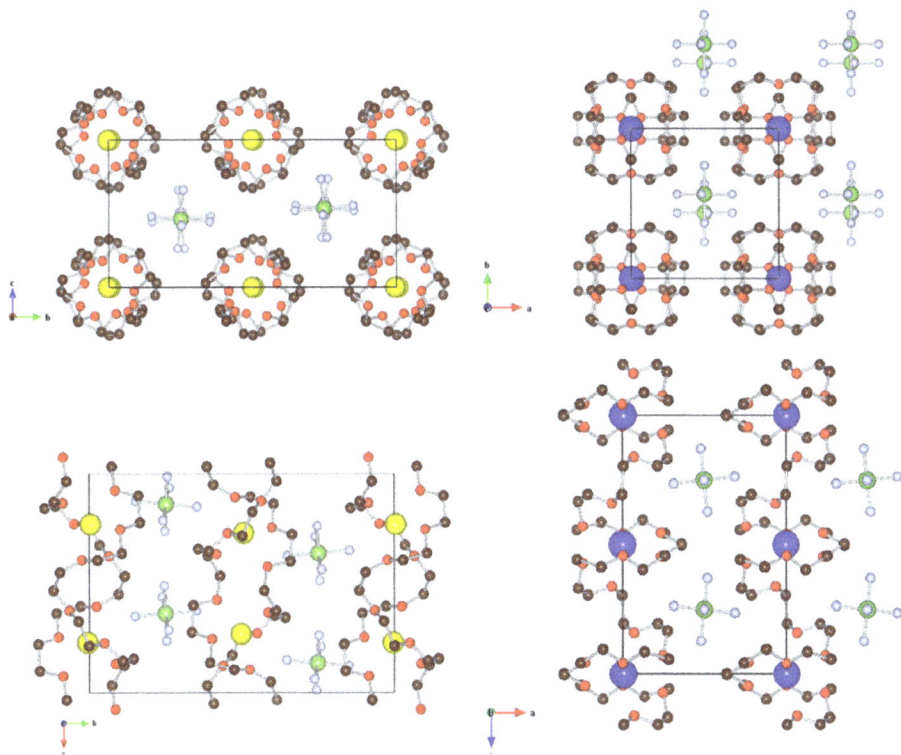

Fig. 2.12: Structure of the crystalline complexes PEO$_6$:LiAsF$_6$ (left) and PEO$_8$:NaAsF$_6$ (right), showing the helical wrapping of the Li$^+$ cations by the polymer chains with the anions located outside of the helical structures. Reprinted from [56], Copyright 2011, with permission from Elsevier.

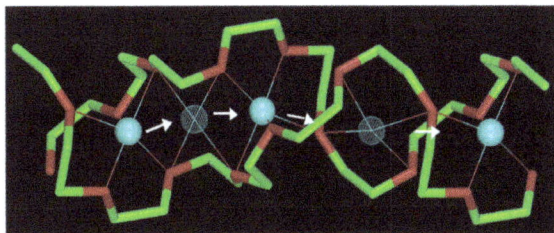

Fig. 2.13: Illustration of the cation transport pathways in the crystalline electrolyte complex PEO$_6$: LiPF$_6$. Reprinted with permission from [53]. Copyright 2003 American Chemical Society.

References

[1]   Cznotka E, Jeschke S, Grünebaum M, Wiemhöfer H-D. Highly-fluorous pyrazolide-based
       lithium salt in PVDF-HFP as solid polymer electrolyte. Solid State Ionics. 2016;292:45–51.
[2]   Sun B, Mindemark J,V, Morozov E, Costa LT, Bergman M, Johansson P, et al. Ion transport in
       polycarbonate based solid polymer electrolytes: Experimental and computational
       investigations. Phys Chem Chem Phys. 2016;18:9504–13.
[3]   Johansson P. First principles modelling of amorphous polymer electrolytes: Li+–PEO, Li+–PEI,
       and Li+–PES complexes. Polymer. 2001;42:4367–73.
[4]   Clancy S, Shriver DF, Ochrymowycz LA. Preparation and characterization of polymeric solid
       electrolytes from poly(alkylene sulfides) and silver salts. Macromolecules. 1986;19:606–11.
[5]   Ionescu-Vasii LL, Garcia B, Armand M. Conductivities of electrolytes based on PEI-b-PEO-b-PEI
       triblock copolymers with lithium and copper TFSI salts. Solid State Ionics. 2006;177:885–92.
[6]   Linert W, Camard A, Armand M, Michot C. Anions of low Lewis basicity for ionic solid state
       electrolytes. Coord Chem Rev. 2002;226:137–41.
[7]   Fenton DE, Parker JM, Wright PV. Complexes of alkali metal ions with poly(ethylene oxide).
       Polymer. 1973;14:589.
[8]   Ratner MA, Shriver DF. Ion transport in solvent-free polymers. Chem Rev. 1988;88:109–24.
[9]   Xu K. Electrolytes and interphases in Li-Ion batteries and beyond. Chem Rev. 2014;114:
       11503–618.
[10]   Mogensen R, Colbin S, Younesi R. An attempt to formulate non-carbonate electrolytes for
       sodium-ion batteries. Batteries & Supercaps. Doi: 10.1002/batt.202000252.
[11]   Gutmann V. Empirical parameters for donor and acceptor properties of solvents. Electrochim
       Acta. 1976;21:661–70.
[12]   Armand M. The history of polymer electrolytes. Solid State Ionics. 1994;69:309–19.
[13]   Åvall G, Mindemark J, Brandell D, Johansson P. Sodium-ion battery electrolytes: Modeling and
       simulations. Adv Energy Mater. 2018;8:1703036.
[14]   Angell CA, Liu C, Sanchez E. A new type of cation-conducting rubbery solid electrolyte:
       The ionic rubber. MRS Proc. 1992;293:75.
[15]   Fan J, Marzke RF, Angeill CA. Conductivity vs. Nmr correlation times, and decoupled cation
       motion in polymer-in-salt electrolytes. MRS Proc. 1992;293:87.
[16]   Austen Angell C, Fan J, Liu C, Lu Q, Sanchez E, Xu K. Li-conducting ionic rubbers for lithium
       battery and other applications. Solid State Ionics. 1994;69:343–53.
[17]   Forsyth M, Jiazeng S, MacFarlane DR. Novel high salt content polymer electrolytes based on
       high Tg polymers. Electrochim Acta. 2000;45:1249–54.
[18]   Forsyth M, Sun J, Macfarlane DR, Hill AJ. Compositional dependence of free volume in PAN/
       LiCF3SO3 polymer-in-salt electrolytes and the effect on ionic conductivity. J Polym Sci Part B:
       Polym Phys. 2000;38:341–50.
[19]   Gerz I, Lindh EM, Thordarson P, Edman L, Kullgren J, Mindemark J. Oligomer electrolytes for
       light-emitting electrochemical cells: Influence of the end groups on ion coordination, ion
       binding, and turn-on kinetics. ACS Appl Mater Interfaces. 2019;11:40372–81.
[20]   Mindemark J, Lacey MJ, Bowden T, Brandell D. Beyond PEO – alternative host materials for
       Li+-conducting solid polymer electrolytes. Prog Polym Sci. 2018;81:114–43.
[21]   Gouverneur M, Schmidt F, Schönhoff M. Negative effective Li transference numbers in Li salt/
       ionic liquid mixtures: Does Li drift in the "Wrong" direction? Phys Chem Chem Phys.
       2018;20:7470–8.
[22]   Baril D, Michot C, Armand M. Electrochemistry of liquids vs. solids: Polymer electrolytes.
       Solid State Ionics. 1997;94:35–47.

[23] Zhang B, Tan R, Yang L, Zheng J, Zhang K, Mo S, et al. Mechanisms and properties of ion-transport in inorganic solid electrolytes. Energy Storage Mater. 2018;10:139–59.

[24] Berthier C, Gorecki W, Minier M, Armand MB, Chabagno JM, Rigaud P. Microscopic investigation of ionic conductivity in alkali metal salts-poly(ethylene oxide) adducts. Solid State Ionics. 1983;11:91–5.

[25] Druger SD, Nitzan A, Ratner MA. Dynamic bond percolation theory: A microscopic model for diffusion in dynamically disordered systems. I. Definition and one-dimensional case. J Chem Phys 1983;79:3133–42.

[26] Devaux D, Bouchet R, Glé D, Denoyel R. Mechanism of ion transport in PEO/LiTFSI complexes: Effect of temperature, molecular weight and end groups. Solid State Ionics. 2012;227:119–27.

[27] de Gennes PG. Reptation of a polymer chain in the presence of fixed obstacles. J Chem Phys. 1971;55:572–9.

[28] Shi J, Vincent CA. The effect of molecular weight on cation mobility in polymer electrolytes. Solid State Ionics. 1993;60:11–17.

[29] Walden P. Über organische Lösungs- und Ionisierungsmittel. Zeitschrift für Physikalische Chemie. 1906;55U:207.

[30] Wang Y, Fan F, Agapov AL, Yu X, Hong K, Mays J, et al. Design of superionic polymers – New insights from Walden plot analysis. Solid State Ionics. 2014;262:782–4.

[31] White RP, Lipson JEG. Polymer free volume and its connection to the glass transition. Macromolecules. 2016;49:3987–4007.

[32] Vogel H. The temperature dependence law of the viscosity of fluids. Physikalische Zeitschrift. 1921;22:645–6.

[33] Fulcher GS. Analysis of recent measurements of the viscosity of glasses. J Am Ceram Soc. 1925;8:339–55.

[34] Tammann G, Hesse W. Die Abhängigkeit der Viscosität von der Temperatur bie unterkühlten Flüssigkeiten. Zeitschrift für anorganische und allgemeine Chemie. 1926;156:245–57.

[35] Angell CA. Polymer electrolytes – Some principles, cautions, and new practices. Electrochim Acta. 2017;250:368–75.

[36] Bruce PG, Gray FM. Polymer electrolytes II: Physical principles. In: Bruce PG, ed. Solid state electrochemistry. Cambridge, Cambridge University Press, 1995, 119–62.

[37] Williams ML, Landel RF, Ferry JD. The temperature dependence of relaxation mechanisms in amorphous polymers and other glass-forming liquids. J Am Chem Soc. 1955;77:3701–7.

[38] Tominaga Y, Yamazaki K. Fast Li-ion conduction in poly(ethylene carbonate)-based electrolytes and composites filled with TiO2 nanoparticles. Chem Commun. 2014;50:4448–50.

[39] Mishra R, Baskaran N, Ramakrishnan PA, Rao KJ. Lithium ion conduction in extreme polymer in salt regime. Solid State Ionics. 1998;112:261–73.

[40] Borodin O, Smith GD, Henderson W. Li+ Cation environment, transport, and mechanical properties of the LiTFSI Doped N-Methyl-N-alkylpyrrolidinium+TFSI- ionic liquids. J Phys Chem B. 2006;110:16879–86.

[41] Li J, Lin Y, Yao H, Yuan C, Liu J. Tuning thin-film electrolyte for lithium battery by grafting cyclic carbonate and combed poly(ethylene oxide) on polysiloxane. ChemSusChem. 2014;7:1901–8.

[42] Kunze M, Karatas Y, Wiemhöfer H-d, Eckert H, Schönhoff M. Activation of transport and local dynamics in polysiloxane-based salt-in-polymer electrolytes: A multinuclear NMR study. Phys Chem Chem Phys. 2010;12:6844–51.

[43] Zhang ZC, Jin JJ, Bautista F, Lyons LJ, Shariatzadeh N, Sherlock D, et al. Ion conductive characteristics of cross-linked network polysiloxane-based solid polymer electrolytes. Solid State Ionics. 2004;170:233–8.

[44] Blonsky PM, Shriver DF, Austin P, Allcock HR. Polyphosphazene solid electrolytes. J Am Chem Soc. 1984;106:6854–5.

[45] Mindemark J, Imholt L, Brandell D. Synthesis of high molecular flexibility polycarbonates for solid polymer electrolytes. Electrochim Acta. 2015;175:247–53.

[46] Pesko DM, Webb MA, Jung Y, Zheng Q, Miller TF, Coates GW, et al. Universal relationship between conductivity and solvation-site connectivity in ether-based polymer electrolytes. Macromolecules. 2016;49:5244–55.

[47] Ebadi M, Eriksson T, Mandal P, Costa LT, Araujo CM, Mindemark J, et al. Restricted ion transport by plasticizing side chains in polycarbonate-based solid electrolytes. Macromolecules. 2020;53:764–74.

[48] Rosenwinkel MP, Andersson R, Mindemark J, Schönhoff M. Coordination effects in polymer electrolytes: Fast Li+ transport by weak ion binding. J Phys Chem C. 2020;124:23588–96.

[49] Angell CA. Fast ion motion in glassy and amorphous materials. Solid State Ionics. 1983; 9–10:3–16.

[50] Angell CA. Mobile ions in amorphous solids. Annu Rev Phys Chem. 1992;43:693–717.

[51] McLin MG, Angell CA. Frequency-dependent conductivity, relaxation times, and the conductivity/viscosity coupling problem, in polymer-electrolyte solutions: LiClO4 and NaCF3SO3 in PPO 4000. Solid State Ionics. 1992;53–56:1027–36.

[52] Gadjourova Z, Andreev YG, Tunstall DP, Bruce PG. Ionic conductivity in crystalline polymer electrolytes. Nature. 2001;412:520–3.

[53] Stoeva Z, Martin-Litas I, Staunton E, Andreev YG, Bruce PG. Ionic conductivity in the crystalline polymer electrolytes PEO6: LiXF6,X = P, As, Sb. J Am Chem Soc. 2003;125: 4619–26.

[54] Christie AM, Lilley SJ, Staunton E, Andreev YG, Bruce PG. Increasing the conductivity of crystalline polymer electrolytes. Nature. 2005;433:50–3.

[55] Zhang C, Gamble S, Ainsworth D, Slawin AMZ, Andreev YG, Bruce PG. Alkali metal crystalline polymer electrolytes. Nat Mater. 2009;8:580–4.

[56] Liivat A. New crystalline NaAsF6–PEO8 complex: A Density Functional Theory study. Electrochim Acta. 2011;57:244–9.

3 Key metrics and how to determine them

The usefulness of an SPE material cannot be determined by a single property; instead, the electrolyte faces several parallel requirements to ascertain its long-term performance and functionality. This chapter will give an overview of factors that determine the usefulness of SPEs and what measurement techniques that are suitable for assessing them. The purpose of the chapter is not to provide a comprehensive theoretical treatment of these techniques, but rather to point toward their practical application for the investigation of SPEs, particularly highlighting peculiarities that apply specifically to solid polymeric systems.

3.1 Total ionic conductivity

Seeing as how the main role of the electrolyte is to transport ions in electrochemical systems, the ionic conductivity is the most crucial property to determine. The total ionic conductivity σ_{tot}, that is, the sum of the conductivity of all ionic species in the system, is determined from the bulk ionic resistance R_b of an electrolyte sample and calculated using knowledge of the thickness l and area A of the sample according to

$$\sigma_{tot} = \frac{l}{R_b \cdot A} \tag{3.1}$$

In contrast with liquid electrolytes, where the geometric parameters of the measurement cell are typically calibrated using a standard of known conductivity, measurements on solid electrolytes are dependent on accurate determination of l and A. Of these, particularly l can be tricky both to measure and to ensure that it remains constant throughout the conductivity measurement. This increases the complexity of the measurement and requires more caution in the interpretation of the data.

The bulk ionic resistance is generally determined using electrochemical impedance spectroscopy (EIS). The measurement is usually done using a two-electrode setup, sandwiching the electrolyte between blocking electrodes (Fig. 3.1). Non-blocking electrodes can also be used, although the data interpretation becomes slightly different depending on the type of electrode used. In the following description, only blocking electrodes will be considered. For interpretation of spectra from symmetrical non-blocking systems, please refer to Section 3.2 on transference number measurements.

A typical blocking electrode setup is stainless steel | SPE | stainless steel and the measurement is performed around OCV with a small voltage amplitude of ca. 10 mV. The obtained data can either be represented in a Bode plot (impedance and phase angle vs frequency) or – more commonly – in a Nyquist plot ($-Z''$ vs Z'). The bulk ionic resistance is found at the plateau in the real part of the impedance, which coincides with a minimum in phase angle, and can reliably be extracted from either type of plot.

https://doi.org/10.1515/9781501521140-003

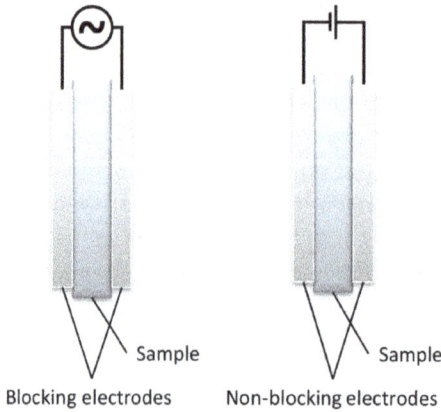

Blocking electrodes   Non-blocking electrodes

Fig. 3.1: Two-electrode setup for EIS measurements of ionic conductivity of SPEs (left) and for polarization experiments (right).

The typical Nyquist plot of an SPE features a semicircle at high frequencies followed by a vertical or near-vertical tail at low frequencies, as illustrated in Fig. 3.2. The semicircle is a result of the geometric capacitance, as the two-electrode cell essentially forms a parallel-plate capacitor with the electrolyte as a dielectric medium. This distinguishes SPEs from liquid electrolytes, where the semicircle is typically *not* observed, as the higher ionic conductivity of liquids shifts it toward frequencies exceeding the capability of standard measurement equipment. Similarly, the semicircle will shift toward higher frequencies as the ionic conductivity increases at elevated temperatures, and only the tail is therefore often observed at high temperatures.

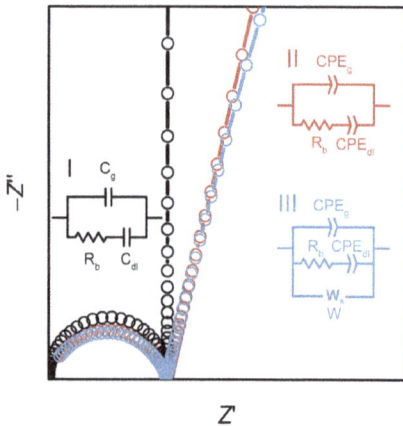

Fig. 3.2: Examples of Nyquist plots corresponding to three different equivalent circuits with relevance for the EIS analysis of polymer electrolytes in a cell with blocking electrodes. C_g/CPE_g correspond to the geometric capacitance, C_{dl}/CPE_{dl} to the double-layer capacitance, R_b to the bulk ionic resistance and W to the Warburg (short) impedance due to ion diffusion in the electrolyte.

The bulk ionic resistance R_b can either be determined as the impedance at the low-frequency intersection of the semicircle with the real axis (Fig. 3.3) or through fitting of the data to an appropriate equivalent circuit. Figure 3.2 compares the impedance response of three slightly different, but physically relevant, equivalent circuits for SPEs. While all these circuits may be useful for extracting R_b with high accuracy, *constant phase elements* (CPEs), representing the effects of imperfect capacitors, better represent the effects of real electrode surfaces than ideal capacitors do. As seen in circuit II, this results in a depression of the semicircle and a slight angle of the low-frequency tail. Accounting for the ionic diffusion in the electrolyte through a Warburg element (circuit III) can additionally provide a better fit at the lowest frequencies, but makes little difference for the extraction of R_b. It should be noted that the data in Fig. 3.3 shows a deviation at high frequencies from the response of the equivalent circuit (see Fig. 3.2), seen as a spiraling inward that is reaching below the real axis. This should typically be interpreted as a high-frequency artifact caused by stray capacitances. For this reason, it is rarely useful to measure at frequencies above 1 MHz. By discarding the highest-frequency data points and instead applying the fitting starting from the top of the semicircle, a reliable fit to the data can nevertheless often be obtained.

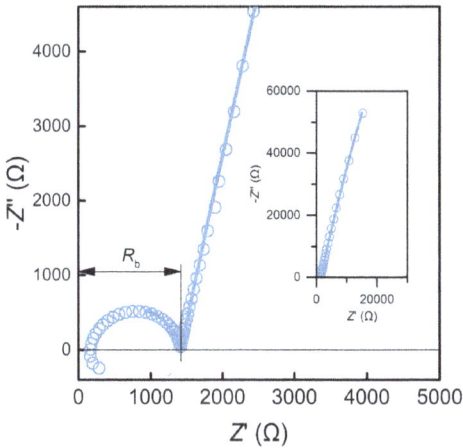

Fig. 3.3: Nyquist plot of EIS data from a poly(ε-caprolactone-*co*-trimethylene carbonate):LiTFSI electrolyte (open circles) that has been fitted to circuit III in Fig. 3.2 (solid line). The inset shows the agreement of the fit in the extended low-frequency tail.

3.2 Transference and transport numbers

Since it is generally either the cation or the anion that is of relevance for the function of a battery, the total ionic conductivity is only of limited relevance to predict the performance of an SPE. Instead, the conductivity of the specific species of interest is given by

$$\sigma_i = t_i \cdot \sigma_{tot} \tag{3.2}$$

This means that the transport (or transference) number needs to be determined as well. A low transference number of the relevant ion leads to the formation of large concentration gradients during operation of the battery cell. This may lead to effects such as salt depletion or precipitation that are detrimental to the operation of the battery.

As stated in Chapter 2, there exists a long-standing confusion about the distinction between transport and transference numbers [1]. Since SPEs typically have salt concentrations far from the dilute limit where $T_i = t_i$, and since most measurement techniques cannot distinguish between contributions from the free ions relative to clustered species, T_i will be the more relevant parameter. Furthermore, the most relevant battery chemistries for SPEs are cation-based (Li^+, Na^+, etc.); hence the discussion will focus on the determination of T_+, particularly for lithium systems. Figure 3.4 compares reported values of T_+ for PEO:LiTFSI electrolytes, which is the most well-studied SPE system, determined using a variety of different methods. As is obvious from the summary in Fig. 3.4, appropriate determination of T_+ is far from straightforward.

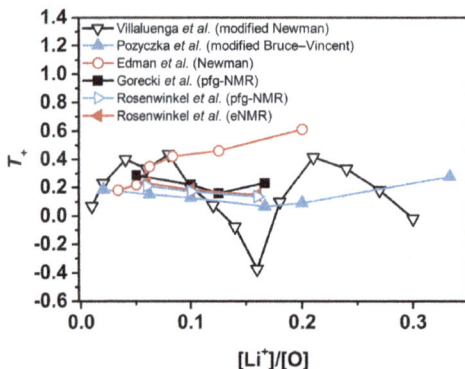

Fig. 3.4: Comparison of the cation transference number for PEO:LiTFSI electrolytes as determined by a variety of methods. Error bars have been omitted for clarity. Data from [2–6].

In the dilute limit, the transference number is essentially a comparison of the relative mobilities of the ions. For a simple binary salt, the cation transference number is calculated as

$$T_+ = \frac{\mu_+}{\mu_+ + \mu_-} \tag{3.3}$$

Since the ionic mobility is proportional to the diffusion coefficient, as given by the Einstein relation in Equation (2.12), the transference number can under these conditions be calculated from self-diffusion coefficients determined from pulsed field gradient (pfg) NMR:

$$T_+ = \frac{D_+}{D_+ + D_-} \tag{3.4}$$

This can be conveniently done for lithium salts with fluorine-containing anions, since ^6Li, ^7Li and ^{19}F are all NMR-active, but may be considerably more difficult if no suitable nuclei are present in the respective ions. As already stated, the assumption of fully dissociated salts does not hold in most practically relevant systems, and any transference number determined by pfg-NMR will incorrectly contain contributions also from neutral ion pairs.

In lithium systems, T_+ can be determined electrochemically through potentiostatic polarization of a symmetric Li | SPE | Li cell (Fig. 3.1) combined with EIS measurements. The most popular version of this method, the Bruce–Vincent method [7, 8], combines polarization at a relatively low voltage $\Delta V = 10$ mV with EIS to determine the interfacial resistances of the Li/SPE interfaces before polarization ($R_{int,0}$) and at steady-state ($R_{int,ss}$), as shown in Fig. 3.5. The determined resistance values are then used to correct the ratio of the steady-state current I_{ss} and initial current I_0 to calculate T_+ for a simple binary salt:

$$T_+ = \frac{I_{ss}(\Delta V - I_0 R_{int,0})}{I_0(\Delta V - I_{ss} R_{int,ss})} \tag{3.5}$$

Some issues with the Bruce–Vincent method include the accurate determination of the interfacial resistances, which can have a large impact on the final calculated value for the transference number. It is also not completely straightforward to determine I_0, as the initial current response is largely capacitive and the exact value recorded for the first data point depends on the response rate of the instrument. This can be circumvented by extracting I_0 from an inverse Cottrell plot, as illustrated in Fig. 3.6. The initial current can also be calculated using Ohm's law according to the approach by Hiller et al. [9]:

$$I_0 = \frac{\Delta V}{R_b + R_{int,\,0}} \tag{3.6}$$

The required bulk and interfacial resistance values can be conveniently determined from the EIS measurements performed before polarization.

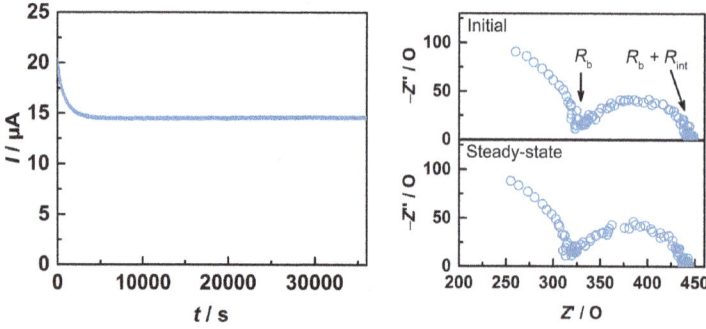

Fig. 3.5: Determination of T_+ through potentiostatic polarization of a Li | SPE | Li cell (left) and determination of the interfacial resistance R_{int} through EIS (right). Data from a poly(ε-caprolactone-co-trimethylene carbonate):LiTFSI electrolyte at 60 °C.

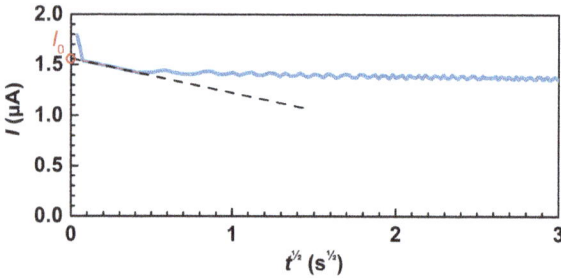

Fig. 3.6: Determination of I_0 during potentiostatic polarization from an inverse Cottrell plot. I_0 is the intersection with the y-axis of the extrapolated linear region at the beginning of polarization. Data from a poly(ε-caprolactone-co-trimethylene carbonate):NaFSI sample.

Another variant of the potentiostatic polarization was proposed by Watanabe et al. [10] where T_+ is calculated using the bulk and interfacial resistances together with the steady-state polarization current according to

$$T_+ = \frac{R_b}{\frac{\Delta V}{I_{SS}} - R_{int}} \tag{3.7}$$

Recent data shows good agreement between the Bruce–Vincent and Watanabe methods [3]. With a symmetric Li | SPE | Li cell, it is also possible to determine T_+ from only impedance measurements without the potentiostatic polarization step, using an approach by Sørensen and Jacobsen [11] and utilizing the impedance response at very low frequencies $Z_d(0)$ according to

$$T_+ = \frac{1}{1 + Z_d(0)/R_b} \tag{3.8}$$

While experimentally straightforward to implement, these electrochemical methods all rely on one key assumption: full dissociation of the salt into free cations and anions. As has already been established, this assumption generally does not hold. The Bruce–Vincent method has, owing to its relative simplicity, nevertheless become a de facto standard method that is widely used. Although the apparent transference numbers attained this way do not necessarily represent the true transport or transference numbers, they still contain useful information about the ion transport characteristics of the system. The near-universal adoption of the Bruce–Vincent method also facilitates reasonable comparisons between different systems, despite the method's shortcomings.

Moreover, the accurate determination of transference numbers through potentiostatic polarization and other electrochemical methods requires electrodes that enable facile stripping/plating of the relevant metal at a stable potential. While this assumption may hold reasonably well for lithium, it is much more uncertain if this is true when trying to measure transference numbers for ions such as Na^+, K^+, Ca^{2+} and Mg^{2+} [12, 13]. The complexity of ion speciation (described in more detail in Section 2.1) in multivalent systems additionally adds further complications [14]. Reported values for these ions should therefore be interpreted with a fair amount of discretion.

Given these limitations, there is understandably an ongoing discussion about the reliability and appropriate measurement practices of transference numbers where several authors have reviewed the limitations of the commonly used methods and the values that have been reported using them, and highlighted the difficulties of fulfilling model assumptions [3, 6]. Methods have also been developed to address, in particular, the problem of assuming full ion dissociation. Newman and coworkers suggested a method based on concentrated solution theory that combines the use of concentration cell, restricted diffusion and current interrupt experiments to calculate T_+ [4, 15]. A modified version of this method was recently presented by Balsara et al., which combines potentiostatic polarization according to the Bruce–Vincent method, restricted diffusion measurements and determination of the "thermodynamic factor" using concentration cells [2, 16]. Since several physicochemical quantities need to be determined using different experimental setups to calculate the transference number with these approaches, one obvious challenge lies in obtaining all values with sufficient precision.

While the methods based on concentrated solution theory aim to amend the flawed assumptions plaguing the frequently applied electrochemical methods, electrophoretic NMR (eNMR) does the same for the traditional pfg-NMR approach. Based on a conventional pfg-NMR experiment, eNMR additionally applies an electric field to the sample, resulting in a correlated drift of charged species that enables accurate determination of the drift velocities of electrically active species containing an NMR-active nucleus. This allows the transference number to be calculated without making prior (and potentially invalid) assumptions about ion speciation. A limitation of this method is that it is limited to relatively low-molecular-weight systems and/or elevated temperatures, where the ionic mobility is sufficiently high. Nevertheless, it has recently been applied to oligomeric analogs of SPE systems [6, 17]. The obtained data appears to correlate well with results obtained using the Bruce–Vincent method, but thus far not with those obtained using the Newman approach (see Fig. 3.4).

Considering the issues connected with the accurate determination of transport numbers, but acknowledging the importance of emphasizing the specific transport of only the relevant ion, it has been suggested as a more practical solution to instead determine the *limiting current (density)* of symmetrical Li ‖ Li cells; that is, the maximum cationic current that can be sustained [18, 19]. This approach circumvents theoretical discussions on what specific information is obtained from the measurements, while at the same time measuring a parameter with direct relevance for battery operation.

3.3 Thermal properties

As the ion transport in SPEs is so closely linked to the dynamics of the polymer chains – at least for the conventional coupled mode of ion transport (see Chapter 2) – it becomes relevant to get an idea of the polymer dynamics for prospective SPE systems. The most directly accessible metric describing the flexibility and dynamics of polymer chains is the glass transition temperature (T_g), which is inversely related to the flexibility of the polymer chains. Generally, a polymer with a low T_g will have faster chain dynamics at a given temperature than a polymer with a higher T_g, although the exact dependence of the segmental dynamics of the system on temperature when approaching T_g may vary between different systems. As such, the T_g can be used for straightforward comparison between different host polymers and SPEs. On dissolution of salt in a host polymer, the T_g will typically increase because of the stiffening effect of the transient cross-links induced by the ion–polymer interactions.

It is not uncommon that ionic conductivity is reported also for high-T_g polymers, and then at temperatures below their T_g value. This implies that the ions are transported through a decoupled mode of transport, either in the solid matrix or facilitated by liquid components remaining in the polymer material after casting. Solvent residues can, for example, give rise to this effect [20–22].

The T_g can be determined by differential scanning calorimetry (DSC), seen as a step in the heat capacity of the material. Mechanical measurements by either rheology or dynamic mechanical analysis (DMA) can also give the T_g as a sudden change in modulus as the temperature is changed. It should be noted, however, that because the glass transition is a second-order transition, it is highly dependent on the experimental conditions, most notably the temperature sweep rate. As such, the measured values may well differ between different measurement techniques. A more thorough treatment of the polymer dynamics involves direct determination of the segmental relaxation time, which can be done through, for example, dielectric spectroscopy or rheology, although this arguably lies beyond the standard repertoire of SPE testing.

DSC can also detect melting and crystallization of semicrystalline systems and measure the heat released or absorbed in these events. As crystallization in polymers is often kinetically limited, the temperature sweep rates become important also for these thermal events. Rapid quenching of a material above the melting point may allow for cooling it below T_g while retaining it in a fully amorphous state. If the heat of melting ΔH_m of a semicrystalline sample is determined from DSC measurements, the degree of crystallinity X_c of the sample prior to melting can be calculated by comparing it to the heat of melting ΔH_m^0 of a (theoretical) 100% crystalline sample:

$$X_c = \left(\frac{\Delta H_m}{\Delta H_m^0 \times (1 - \phi_{add})} \right) \tag{3.9}$$

In this version of the equation, the total amount of additives ϕ_{add} (salt, nanoparticles, etc.) of the electrolyte is accounted for [23].

Since SPEs are often used at elevated temperatures, it may be of relevance to determine the thermal stability at the intended operating temperature. This is conveniently done by thermogravimetric analysis (TGA). Care must be taken that some salts react with aluminum pans at high temperatures. TGA is also difficult to perform under completely inert conditions, which means that moisture absorption may skew the results. The most common practice is to run a temperature ramp under nitrogen flow and detecting the onset of degradation as the onset of significant weight loss. However, this dynamic mode of measurement typically overestimates the thermal stability, and there can be significant thermal degradation – albeit too slow to detect during a temperature ramp – at much lower temperatures. More accurate results will therefore be obtained through isothermal measurements, where the temperature is instead increased stepwise, allowing the detection of even minute levels of weight loss [24].

3.4 Mechanical properties

The coupling of ion transport to polymer chain dynamics also results in an inverse relationship between ion transport and the mechanical properties of the SPE such that a high-conductivity electrolyte will be soft, while a hard electrolyte will have low conductivity. This can to some extent be mitigated by cross-linking, where a microscopic polymer flexibility remains while the material macroscopically behaves like a rubbery solid. Another strategy is the use of block copolymers to separate the mechanical properties in a hard block from the ion transport in a soft block. However, the trade-off in mechanical and conductive properties is inevitable for classical amorphous SPEs, while the semicrystalline counterparts will lose mechanical integrity around T_m. In either case, sufficient mechanical properties are relevant to ascertain safe and stable battery functionality, but is often overlooked or disregarded in the quest for higher ionic conductivities.

Most SPEs are used far above their T_g and show typical viscoelastic properties. If there are no covalent cross-links, semicrystallinity or stabilizing hard blocks, the material will therefore primarily gain its mechanical properties from chain entanglements and transient physical cross-links due to ion–polymer interactions. This can be noticed as an increase in mechanical stability and a transition toward a more solid-like mechanical response as salt is added to the polymer matrix (Fig. 3.7). There may also be notable plasticization by the salt, particularly at high concentrations. This may lead to creep and stress relaxation behavior, which may cause short circuits in a battery, where the electrolyte is subjected to a constant load for an extended time.

Viscoelastic properties can be investigated through rheology (shearing) or DMA (several modes of deformation are possible). By applying an oscillating stress, a phase-shifted strain can be detected and the complex modulus can be determined. In oscillatory rheology, the complex shear modulus G^* will be measured:

$$G^* = G' + iG'' \tag{3.10}$$

where G' is the *storage modulus*, which describes the elastic response, G'' is the *loss modulus*, which describes the viscous response and i is the imaginary unit. For DMA, the shear modulus may be replaced by, for example, the tensile modulus instead. By running a frequency sweep, the viscoelastic response over a frequency range can be determined, and this range can be increased by running measurements at different temperatures and constructing a master curve by applying time–temperature superposition. The response at very low frequencies can be used as an indicator of the response under constant load. If the loss modulus is higher than the storage modulus, the response to deformation will primarily be viscous and the behavior can be described as "liquid-like." As the oscillation frequency increases, there will usually be a crossover to "solid-like" behavior at a certain frequency (as shown in Fig. 3.7) above which the response is instead dominated by the storage modulus. This frequency is highly temperature-dependent, and at higher temperatures the crossover will occur at higher frequencies, indicating a loss of mechanical stability. A covalently cross-linked material will have roughly parallel storage and loss moduli that are independent of frequency.

3.5 Electrochemical stability

For long-term performance in an electrochemical system, the electrolyte needs to be electrochemically stable at the potentials of the redox processes at the electrodes. This does not necessarily mean that the electrolyte needs to be thermodynamically

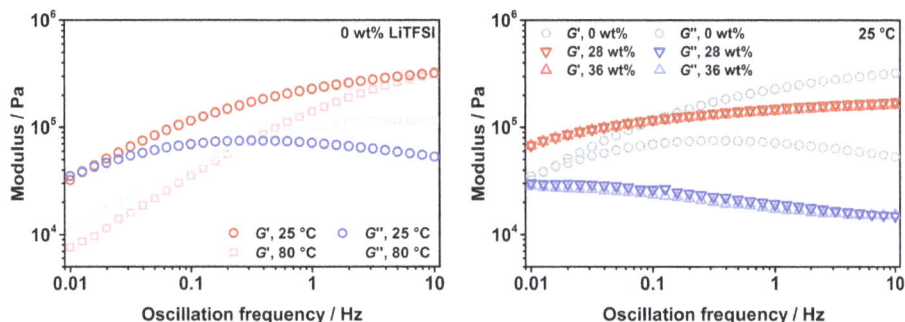

Fig. 3.7: Comparison of the rheological response at different oscillation frequencies for poly (ε-caprolactone-co-trimethylene carbonate) at different temperatures (left) and with incorporation of different concentrations of LiTFSI salt (right). A shift of the crossover to higher frequencies with increased temperature can be seen (left), as well as a shift to lower frequencies with addition of salt (right). This indicates a more solid-like and mechanically stable behavior for the electrolyte compared to the pure polymer. Adapted from [25], Copyright 2017, with permission from Elsevier.

stable – kinetic stability may be sufficient to enable stable operation. As such, it is of relevance to determine the *electrochemical stability window* (ESW) of the electrolyte (Fig. 3.8). A common misconception is that the ESW is determined by the HOMO and LUMO of the electrolyte [26], but as with any chemical process, not only the starting material but also the products need to be considered to assess its thermodynamics. For polymer electrolytes, density functional theory (DFT) modeling has demonstrated that the ESW depends on the specific combination of host material and salt, and is different from the HOMO/LUMO of either of them (Fig. 3.9) [27].

It is widely accepted that with most typical negative electrodes for Li-based batteries (lithiated graphite, Li metal), degradation at this electrode is inevitable because the ESW of the electrolyte is unlikely to overlap with the comparatively extreme redox potential of the negative electrode. Stability at the negative electrode is instead dependent on the formation of effective protective passivation layers such as the SEI layer (see Chapter 1). In SPE-based Li batteries, the *practical* stability at the negative electrode is commonly assessed through Li stripping/plating experiments using a symmetric Li-metal cell, but such experiments might not give much information about the true electrochemical stability at these low potentials. Efforts to determine the electrochemical stability limits are instead focused on the stability at the high potentials of the positive electrode (cathode).

Measurements of the electrochemical stability are most typically performed using either cyclic or linear sweep voltammetry (CV and LSV, respectively). For practical reasons, the standard cell setup uses a combined counter and reference electrode. In lithium systems, this is composed of a Li-metal foil. The working electrode needs to be electrochemically inert in the potential range of interest. This may require the use of different materials for high- and low-potential sweeps, respectively. Examples include

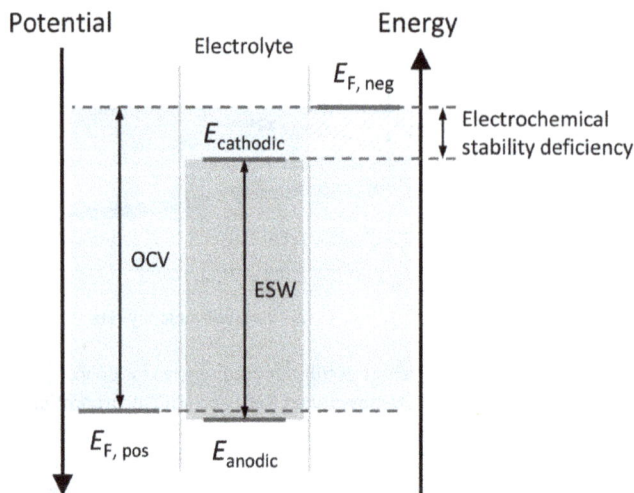

Fig. 3.8: Schematic energy diagram of an electrochemical cell, where there is a gap between the electrochemical stability window (ESW) of the electrolyte and the redox potential of the negative electrode process.

copper, which may oxidize at high potentials but is suitable for the low-potential sweeps, while stainless steel, which already has native surface oxides, may be suitable for the high-potential sweeps.

CV and LSV are relatively straightforward measurement techniques that can be performed on standard electrochemical instruments. The difficulty instead arises when defining the potential limits for the electrochemical stability. Whereas the thermodynamic redox potential can be determined for reversible faradaic reactions at the working electrode, no such clearly defined potential can be found for an irreversible reaction (such as electrochemical electrolyte degradation). Sometimes a cutoff current is defined for the onset of electrochemical degradation, but since the current response is dependent on measurement parameters such as electrode area and scan speed, such a limit will necessarily be arbitrary to some degree. Alternative approaches include looking at the differential current to detect the onset of degradation [28], but there is still the challenge of accurately identifying the true onset of these irreversible reactions. Assertions of electrochemical stability as deduced from voltammetry experiments have furthermore shown poor correlation with the practical stability in real battery systems, and cycling of SPEs versus other negative materials than $LiFePO_4$ (i.e., LCO, NMC, etc.) has proven challenging [28, 29]. The utility of voltammetry experiments for assessing the electrochemical stability of electrolyte systems can therefore be questioned.

Voltammetry approaches – which typically show excellent electrochemical stability of SPE materials – are therefore in fact problematic to use. In addition to the discussion above, this is also partly because of the dynamic conditions of the experiments

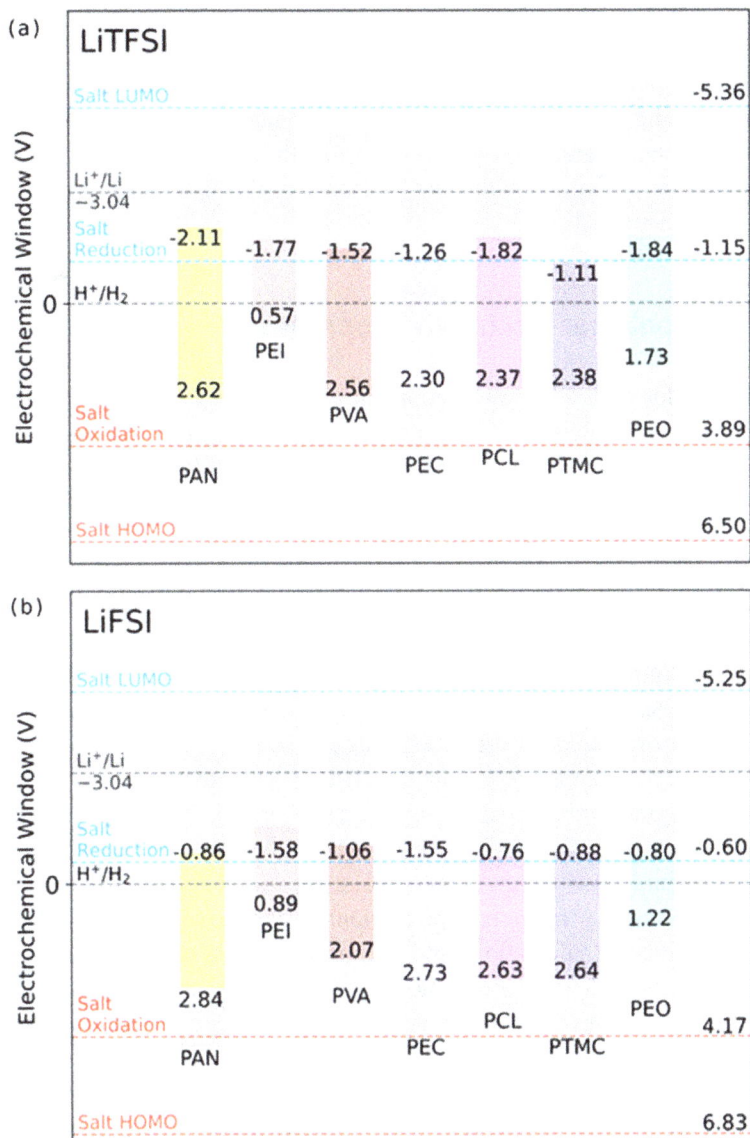

Fig. 3.9: Electrochemical stability of SPEs from DFT calculations compared to the HOMO/LUMO of the salt as well as the redox potential of Li metal. The plot also shows the ESWs of the pristine polymers (shaded bars) and salts (dashed lines) for comparison. Adapted with permission from [27]. Copyright 2020 American Chemical Society.

and partly because they are performed with working electrodes that do not resemble the real battery electrodes as the catalytic effects of different materials surfaces may shift the degradation behavior and significantly affect the observed current response

[28, 30, 31]. An alternative approach that simultaneously addresses both these issues is *electrochemical floating analysis* in relevant battery cell setups, where the potential is increased stepwise, allowing for detection of currents from detrimental side reactions under static conditions [32]. Nevertheless, it is difficult to definitely identify a process such as electrolyte degradation without detecting the degradation products by either spectroscopically probing the interface [33] or detecting gas-phase products through mass spectrometry [34].

3.6 Modeling of polymer electrolyte properties

Modeling and simulation tools are growing in importance to complement experimental studies of polymer electrolytes and their implementation in batteries. Generally, modeling tools are adopted for different time and length scales, which means that the choice of method needs to be specified for the problem or question at hand. Today, large efforts are made into combining these methods in a *multi-scale* framework, where they are intrinsically connected and information is passed between the different size and time domains. For example, methods for reaction kinetics are connected with models for transport processes, which in turn are connected with mesoscale structural rearrangements, and which are ultimately all connected to a simulation of battery performance [35]. With the growth in computer capability, the tools used for computationally analyzing batteries are rapidly becoming good enough to use for problems that seemed too complex just years ago. This has led these techniques to become useful also for prediction, and not only for an increased understanding of the molecular systems or devices.

Generally, when employing computational materials science tools, the more refined the approximations are, the more computationally expensive the simulations become. Electronic structure calculations and chemical reactions are here the most advanced, employing *ab initio* or DFT techniques. The simulated system is then by necessity small or can only be employed for periodic structures. For mass transport, on the other hand, which is key for many electrolyte properties, larger system sizes are necessary to capture the structure–dynamics of the system. Force field methods such as molecular dynamics (MD) or kinetic Monte Carlo (kMC) simulations are therefore frequently employed. MD has long been a method of choice for studies of polymer electrolytes [36]. Here, the atom–atom interactions are described through analytical expressions rather than quantum mechanical equations. If going to larger systems, for example micro-scale structures, coarse-graining of the molecular components is necessary, thereby employing mesoscale modeling techniques. At the battery device level, materials modeling is generally of little use, which means that modeling requires analytical descriptions of the relevant processes (e.g., those described in Section 1.2), often expressed as partial differential equations. To solve these, and to be able to do it for complex and three-dimensional structures, finite element methodology (FEM) is nowadays often employed.

Traditionally, electronic structure calculations have been employed to study coordination chemistry between salts and polymers, and in this context also to predict dissociation energy for different types of salts in polymeric systems. Thereby, it can be predicted what salts that will straightforwardly dissolve in different polymer matrices [37]. Moreover, the preferred coordination structures can be determined, which can shed light on different structures that promote ionic mobility, and can also be used for prediction of strong complexation of the cations in these systems [38]. These calculations can typically be benchmarked toward spectroscopy data, for example FT-IR and Raman.

More recently, similar calculation techniques have been used to explore other vital properties of polymer electrolytes. Through combined MD and DFT studies, the *electronic* conductivity of PEO-based electrolytes has been explored [39]. This is a vital property, considering that it is necessary that the SPE material is more or less an electronic insulator and solely an ionic conductor, at least if scaled down to be very thin – which is a commonly suggested strategy for mitigating the problems associated with poor ionic conductivity. These studies have shown that while the polymer host can be a good insulator, the inclusion of salt can reduce its band-gap in a detrimental way (down to 0.6 eV), which can give rise to current leakage and self-discharge problems. Similar methodologies have also been used for exploring the reactivity of different anions and polymers, both physically on Li-metal surfaces [40] and in gas-phase calculations [41]. This gives insights into which salts and polymers that decompose, into what kind of products, and at what potentials. Screening of the electrochemical properties of different SPE materials using DFT methodology has shown that the salt plays a key role for determining the ESW of the electrolyte (Fig. 3.9), where especially the widely used TFSI and FSI salts are problematic [27].

As stated, MD simulations have been extensively used to explore the interplay between structure and dynamics in polymer electrolytes [42–45]. These studies have until recently almost exclusively focused on PEO-based systems, where several different salts and polymer architectures have been studied. MD simulations create a short "movie" of the dynamics in the simulated system, generally in the nanosecond regime. By statistically analyzing the trajectories of the different atoms and employing correlation functions, decisive structures for ion transfer can be pinpointed, as illustrated in Fig. 3.10 [46]. It is here vital to acknowledge two important restrictions when using MD data to compare with experimental counterparts and make conclusions for the ionic mobility: firstly, it is diffusivity that is extracted from any conventional MD simulations, since these generally employ equilibrium conditions. This is fundamentally different from the nonequilibrium conditions employed in an operating cell or in an electrochemical measurement, where migration might dominate over diffusion (see Chapter 2), but which is more problematic to simulate. Secondly, due to the comparatively slow relaxation of the polymeric solvent, extensive simulation times are required

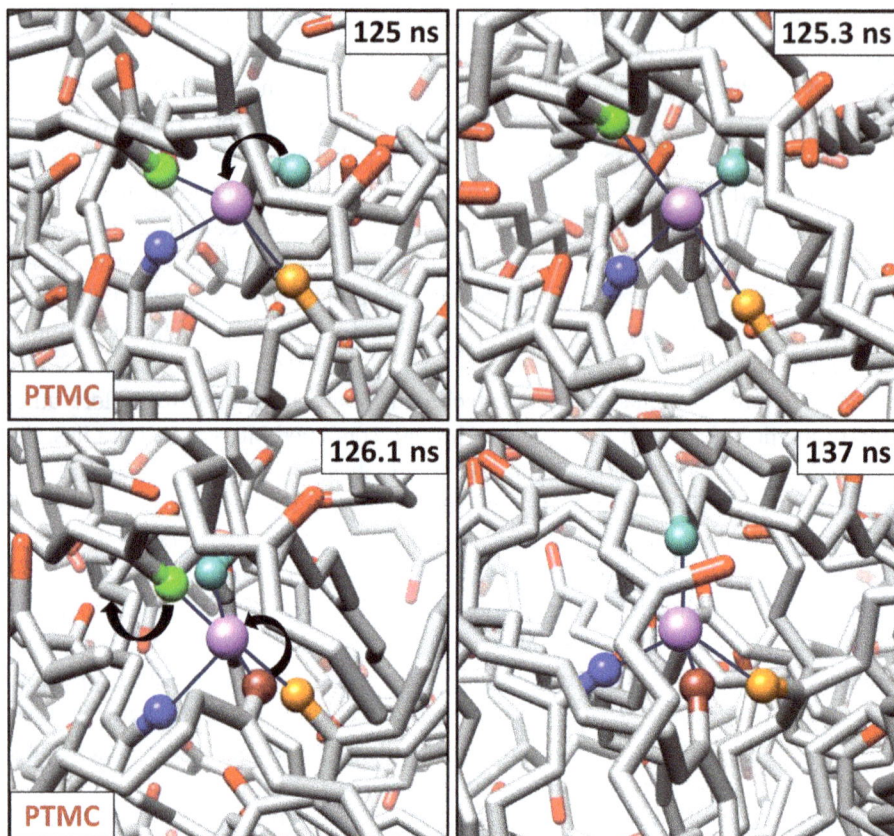

Fig. 3.10: Illustration of changes in the coordination shell of a Li⁺ ion in a poly(trimethylene carbonate) (PTMC) electrolyte with LiTFSI salt from MD trajectories. Carbonyl oxygens have been highlighted (red for non-coordinating moieties and different colors for carbonyl oxygens within 2.5 Å of the Li⁺ cation). Adapted from [46] under CC-BY 4.0 (http://creativecommons.org/licenses/by/4.0/).

to truly capture the diffusive regime in the materials [47]. This means that within the scope of much MD work on SPEs, the material is still in the sub-diffusive domain. Although most transport mechanisms are most likely still adequately captured, their relative importance and quantitative estimations might be off from the corresponding macroscopic system. Thus, despite the domination of MD studies for SPEs, this is not a computational methodology without shortcomings.

For battery applications, also the structure–dynamics properties at the interfacial region between the SPE and the electrode materials are of importance. From the handful of studies that exists on these systems [48–50] – but where chemical reactions of neither salt nor polymer are taken into account, since these are not captured in classical MD studies – it is clear that the presence of the surface leads to reduced polymer dynamics and to accumulation of salt near the surface. This seems to indicate reduced dynamics in this region, which would cause resistances in the battery cell. However, these studies have not yet seriously taken transport *across* this layer into account, and neither have they considered the surface layers (SEI) that will form spontaneously due to the limited ESW of the SPE. These will likely have a profound effect on the true structure–dynamics in the interfacial region.

Mesoscale methods in principle build on coarse-graining, where molecular segments are merged into beads in the model. These rather crude approximations speed up the simulation several orders of magnitude, thereby overcoming many of the problems with low macromolecular mobility and that many relevant structural features appear at the micro-scale. The interactions between the bonded beads are often represented by spring constants, whereas nonbonded interactions are treated by coulombic and Lennard–Jones potentials. These methods have been successfully employed to study for example ionomeric conductors [51] and block-copolymeric SPE systems [52], where local phase separation and percolation are crucial to understanding ion transport phenomena.

While FEM methods are necessary to simulate the entire battery cell, most FEM models of SPE-based batteries have been rather primitive where the polymer electrolyte is simply approximated with a specific ionic conductivity (lower than liquid counterparts) and specific lithium transference number. Other specific parameters of SPEs, for example, pore-filling properties and electrode particle contacts of the electrolyte (see Chapter 1) are commonly neglected. Nevertheless, these results have provided interesting insights into the discharge characteristics and concentration buildup in SPE-based batteries [53]. Moreover, since many FEM tools also employ multiphysics capabilities, the electrochemical description of battery behavior can be intrinsically coupled to the thermal evolution [54] or changing mechanical properties of the polymer [55] during battery operation. With more refined models, and based on input parameters from both experiments and materials modeling, these tools will likely soon become very strong for predicting SPE battery performance.

References

[1]   Baril D, Michot C, Armand M. Electrochemistry of liquids vs. solids: Polymer electrolytes. Solid State Ionics. 1997;94:35–47.

[2]   Villaluenga I, Pesko DM, Timachova K, Feng Z, Newman J, Srinivasan V, et al. Negative Stefan-Maxwell diffusion coefficients and complete electrochemical transport characterization of homopolymer and block copolymer electrolytes. J Electrochem Soc. 2018;165:A2766–A73.

[3]   Pożyczka K, Marzantowicz M, Dygas JR, Krok F. Ionic conductivity and lithium transference number of poly(ethylene oxide): LiTFSI system. Electrochim Acta. 2017;227:127–35.

[4]   Edman L, Doeff MM, Ferry A, Kerr J, De Jonghe LC. Transport properties of the solid polymer electrolyte system P(EO)nLiTFSI. J Phys Chem B. 2000;104:3476–80.

[5]   Gorecki W, Jeannin M, Belorizky E, Roux C, Armand M. Physical properties of solid polymer electrolyte PEO(LiTFSI) complexes. J Phys Condens Matter. 1995;7:6823–32.

[6]   Rosenwinkel MP, Schönhoff M. Lithium transference numbers in PEO/LiTFSA electrolytes determined by electrophoretic NMR. J Electrochem Soc. 2019;166:A1977–A83.

[7]   Evans J, Vincent CA, Bruce PG. Electrochemical measurement of transference numbers in polymer electrolytes. Polymer. 1987;28:2324–8.

[8]   Bruce PG, Vincent CA. Steady state current flow in solid binary electrolyte cells. J Electroanal Chem Interfacial Electrochem. 1987;225:1–17.

[9]   Hiller MM, Joost M, Gores HJ, Passerini S, Wiemhöfer HD. The influence of interface polarization on the determination of lithium transference numbers of salt in polyethylene oxide electrolytes. Electrochim Acta. 2013;114:21–9.

[10] Watanabe M, Nagano S, Sanui K, Ogata N. Estimation of Li+ transport number in polymer electrolytes by the combination of complex impedance and potentiostatic polarization measurements. Solid State Ionics. 1988;28–30:911–7.

[11] Ravn Sørensen P, Jacobsen T. Conductivity, charge transfer and transport number – an ac-investigation of the polymer electrolyte LiSCN-poly(ethyleneoxide). Electrochim Acta. 1982;27:1671–5.

[12] Boschin A, Abdelhamid ME, Johansson P. On the feasibility of sodium metal as pseudo-reference electrode in solid state electrochemical cells. ChemElectroChem. 2017;4:2717–21.

[13] Dugas R, Forero-Saboya JD, Ponrouch A. Methods and protocols for reliable electrochemical testing in post-Li batteries (Na, K, Mg, and Ca). Chem Mater. 2019;31:8613–28.

[14] Park B, Schaefer JL. Review – polymer electrolytes for magnesium batteries: Forging away from analogs of lithium polymer electrolytes and towards the rechargeable magnesium metal polymer battery. J Electrochem Soc. 2020;167:070545.

[15] Ma Y, Doyle M, Fuller TF, Doeff MM, De Jonghe LC, Newman J. The measurement of a complete set of transport properties for a concentrated solid polymer electrolyte solution. J Electrochem Soc. 1995;142:1859–68.

[16] Pesko DM, Timachova K, Bhattacharya R, Smith MC, Villaluenga I, Newman J, et al. Negative transference numbers in poly(ethylene oxide)-based electrolytes. J Electrochem Soc. 2017;164:E3569–E75.

[17] Rosenwinkel MP, Andersson R, Mindemark J, Schönhoff M. Coordination effects in polymer electrolytes: Fast Li+ transport by weak ion binding. J Phys Chem C. 2020;124:23588–96.

[18] Gribble DA, Frenck L, Shah DB, Maslyn JA, Loo WS, Mongcopa KIS, et al. Comparing experimental measurements of limiting current in polymer electrolytes with theoretical predictions. J Electrochem Soc. 2019;166:A3228–A34.

[19] Shah DB, Kim HK, Nguyen HQ, Srinivasan V, Balsara NP. Comparing measurements of limiting current of electrolytes with theoretical predictions up to the solubility limit. J Phys Chem C. 2019;123:23872–81.

[20] MacFarlane DR, Zhou F, Forsyth M. Ion conductivity in amorphous polymer/salt mixtures. Solid State Ionics. 1998;113–115:193–7.

[21] Voigt N, Van Wüllen L. The mechanism of ionic transport in PAN-based solid polymer electrolytes. Solid State Ionics. 2012;208:8–16.

[22] Ek G, Jeschull F, Bowden T, Brandell D. Li-ion batteries using electrolytes based on mixtures of poly(vinyl alcohol) and lithium bis(triflouromethane) sulfonamide salt. Electrochim Acta. 2017;246:208–12.

[23] Eriksson T, Mindemark J, Yue M, Brandell D. Effects of nanoparticle addition to poly(ε-caprolactone) electrolytes: Crystallinity, conductivity and ambient temperature battery cycling. Electrochim Acta. 2019;300:489–96.

[24] Paillard E, Zhou Q, Henderson WA, Appetecchi GB, Montanino M, Passerini S. Electrochemical and physicochemical properties of PY14FSI-Based electrolytes with LiFSI. J Electrochem Soc. 2009;156:A891.

[25] Mindemark J, Sobkowiak A, Oltean G, Brandell D, Gustafsson T. Mechanical stabilization of solid polymer electrolytes through gamma irradiation. Electrochim Acta. 2017;230:189–95.

[26] Peljo P, Girault HH. Electrochemical potential window of battery electrolytes: the HOMO–LUMO misconception. Energy Environ Sci. 2018;11:2306–9.

[27] Marchiori CFN, Carvalho RP, Ebadi M, Brandell D, Araujo CM. Understanding the electrochemical stability window of polymer electrolytes in solid-state batteries from atomic-scale modeling: The role of Li-Ion salts. Chem Mater. 2020;32:7237–46.

[28] Qiu J, Liu X, Chen R, Li Q, Wang Y, Chen P, et al. Enabling stable cycling of 4.2 V high-voltage all-solid-state batteries with PEO-Based solid electrolyte. Adv Funct Mater. 2020;30:1909392.

[29] Homann G, Stolz L, Nair J, Laskovic IC, Winter M, Kasnatscheew J. Poly(Ethylene Oxide)-based electrolyte for solid-state-lithium-batteries with high voltage positive electrodes: evaluating the role of electrolyte oxidation in rapid cell failure. Sci Rep. 2020;10:4390.

[30] Westman K, Dugas R, Jankowski P, Wieczorek W, Gachot G, Morcrette M, et al. Diglyme based electrolytes for sodium-ion batteries. ACS Appl Energy Mater. 2018;1:2671–80.

[31] Dewald GF, Ohno S, Kraft MA, Koerver R, Till P, Vargas-Barbosa NM, et al. Experimental assessment of the practical oxidative stability of lithium thiophosphate solid electrolytes. Chem Mater. 2019;31:8328–37.

[32] Zhao C-Z, Zhao Q, Liu X, Zheng J, Stalin S, Zhang Q, et al. Rechargeable lithium metal batteries with an in-built solid-state polymer electrolyte and a high voltage/loading Ni-rich layered cathode. Adv Mater. 2020;32:1905629.

[33] Sångeland C, Mindemark J, Younesi R, Brandell D. Probing the interfacial chemistry of solid-state lithium batteries. Solid State Ionics. 2019;343:115068.

[34] Li J, Dong S, Wang C, Hu Z, Zhang Z, Zhang H, et al. A study on the interfacial stability of the cathode/polycarbonate interface: implication of overcharge and transition metal redox. J Mater Chem A. 2018;6:11846–52.

[35] Franco AA, Rucci A, Brandell D, Frayret C, Gaberscek M, Jankowski P, et al. Boosting rechargeable batteries R&D by multiscale modeling: Myth or reality? Chem Rev. 2019;119: 4569–627.

[36] Johansson P. Computational modelling of polymer electrolytes: What do 30 years of research efforts provide us today? Electrochim Acta. 2015;175:42–6.

[37] Johansson P. Electronic structure calculations on lithium battery electrolyte salts. Phys Chem Chem Phys. 2007;9:1493–8.

[38] Xue S, Liu Y, Li Y, Teeters D, Crunkleton DW, Wang S. Diffusion of lithium ions in amorphous and crystalline Poly(ethylene oxide)3: LiCF3SO3 polymer electrolytes. Electrochim Acta. 2017;235:122–8.

[39] Unge M, Gudla H, Zhang C, Brandell D. Electronic conductivity of polymer electrolytes: electronic charge transport properties of LiTFSI-doped PEO. Phys Chem Chem Phys. 2020;22:7680–4.

[40] Ebadi M, Marchiori C, Mindemark J, Brandell D, Araujo CM. Assessing structure and stability of polymer/lithium-metal interfaces from first-principles calculations. J Mater Chem A. 2019;7:8394–404.

[41] Scheers J, Pitawala J, Thebault F, Kim J-K, Ahn J-H, Matic A, et al. Ionic liquids and oligomer electrolytes based on the B(CN)4– anion; ion association, physical and electrochemical properties. Phys Chem Chem Phys. 2011;13:14953–9.

[42] Borodin O, Smith GD. Molecular dynamics simulations of poly(ethylene oxide)/LiI Melts. 1. structural and conformational properties. Macromolecules. 1998;31:8396–406.

[43] Borodin O, Smith GD. Molecular dynamics simulations of Poly(ethylene oxide)/LiI Melts. 2. dynamic properties. Macromolecules. 2000;33:2273–83.

[44] Karo J, Brandell D. A molecular dynamics study of the influence of side-chain length and spacing on lithium mobility in non-crystalline LiPF6·PEOx; x=10 and 30. Solid State Ionics. 2009;180:1272–84.

[45] Balbuena PB, Lamas EJ, Wang Y. Molecular modeling studies of polymer electrolytes for power sources. Electrochim Acta. 2005;50:3788–95.

[46] Ebadi M, Eriksson T, Mandal P, Costa LT, Araujo CM, Mindemark J, et al. Restricted ion transport by plasticizing side chains in polycarbonate-based solid electrolytes. Macromolecules. 2020;53:764–74.

[47] Borodin O, Smith GD. Mechanism of ion transport in amorphous poly(ethylene oxide)/LiTFSI from molecular dynamics simulations. Macromolecules. 2006;39:1620–9.

[48] Borodin O, Smith GD, Bandyopadhyaya R, Byutner O. Molecular dynamics study of the influence of solid interfaces on poly(ethylene oxide) structure and dynamics. Macromolecules. 2003;36:7873–83.

[49] Ebadi M, Costa LT, Araujo CM, Brandell D. Modelling the polymer electrolyte/Li-metal interface by molecular dynamics simulations. Electrochim Acta. 2017;234:43–51.

[50] Verners O, Lyulin AV, Simone A. Salt concentration dependence of the mechanical properties of LiPF6/poly(propylene glycol) acrylate electrolyte at a graphitic carbon interface: A reactive molecular dynamics study. J Polym Sci Part B: Polym Phys. 2018;56:718–30.

[51] Lu K, Maranas JK, Milner ST. Ion-mediated charge transport in ionomeric electrolytes. Soft Matter. 2016;12:3943–54.

[52] Qin J, De Pablo JJ. Ordering transition in salt-doped diblock copolymers. Macromolecules. 2016;49:3630–8.

[53] Zadin V, Brandell D. Modelling polymer electrolytes for 3D-microbatteries using finite element analysis. Electrochim Acta. 2011;57:237–43.

[54] Priimagi P, Kasemagi H, Aabloo A, Brandell D, Zadin V. Thermal simulations of polymer electrolyte 3D Li-microbatteries. Electrochim Acta. 2017;244:129–38.

[55] Grazioli D, Zadin V, Brandell D, Simone A. Electrochemical-mechanical modeling of solid polymer electrolytes: Stress development and non-uniform electric current density in trench geometry microbatteries. Electrochim Acta. 2019;296:1142–62.

4 Batteries based on solid polymer electrolytes

4.1 Battery testing

When employing the SPE material into a battery cell, a range of battery testing and analysis techniques become relevant, apart from those which especially concern the electrolyte properties, and which are covered in Chapter 3. Most of these are testing, which also applies to other types of battery systems, containing liquid or ceramic electrolytes, but where the specific properties of polymer electrolytes render a certain typical behavior for SPE-based batteries. It should be noted that it is primarily the last 10 years or so that have seen an emerging regular testing and evaluation of SPE materials in real battery cells; before that, the bulk part of SPE studies mainly concerned extracting electrolyte parameters, and less was known or investigated about their behavior in practical electrochemical devices. Moreover, much of the previously employed SPE-based cell testing often utilized electrodes being wetted with some small amount of liquid electrolyte in order to facilitate ion transport from the electrode to the SPE. While such a strategy can be effective in the short term, it is impossible to control whether or not this liquid is decomposing on the electrode or diffusing into the SPE material, thereby giving a less appropriate impression of the long-term electrochemical stability or transport properties of the system.

It should also be acknowledged that cell testing renders several additional dimensions to the analysis of SPEs, and it can often be difficult to estimate what material-specific properties that are the cause of the observed battery behavior. For example, if cell failure occurs, is it due to bulk electrolyte properties, or incompatibility with the employed anode? Or the cathode? Or due to other failure mechanisms involving salt degradation, corrosion, etc.? While battery testing is necessary to truly capture the functionality of the material, it is usually not a robust or precise method to understand the fundamentals of SPE materials.

Assembling the battery test cell using an SPE material requires a bit more effort than the conventional approach of merely adding liquid electrolyte to a prefabricated battery pouch or coin cell before testing. The polymer electrolyte is, at least for lab-scale batteries, usually fabricated by dissolving the salt and polymer in a common solvent, which is then evaporated to form a homogeneous film [1, 2]. Alternatively, but less commonly, hot-pressing can be employed – thereby ensuring that the film is completely solvent-free [3–5]. The major challenge when constructing an SPE-based battery is to both get a good and stable polymer film, while also achieving a good wetting of the active material in the electrode. The film normally becomes of better quality if cast on a substrate such as a PTFE mold, but if this prefabricated film is then applied onto a porous cathode, it might have a very limited amount of contact points. Casting directly onto a prefabricated electrode is

https://doi.org/10.1515/9781501521140-004

thereby often a better strategy, but can result in a less uniform SPE film, and also problems of removing all solvent from the porous electrode.

Another factor of importance is the evaporation pressure during SPE fabrication. Reduced pressure leads to a more rapid process, but can also cause bubbly and inhomogeneous films. Casting under partially and sequentially reduced pressure is a possible solution to this. Since many polymers and salts are notoriously hygroscopic, casting should be done in inert atmosphere such as N_2 or Ar. Irrespective, it is vital that the gases are dry and firmly controlled, since water contaminants have a tendency to accumulate in the SPE samples, not least for the hygroscopic PEO (poly(ethylene oxide)). Finally, it is also conventional to anneal the battery at a somewhat elevated temperature (60–90 °C) post-assembly. The polymer then softens and its adhesion to the electrodes improves, thereby leading to higher capacity and lower interfacial resistance.

After assembly, the battery *cell testing* is conventionally performed using galvanostatic cycling; that is, employing a constant current and measuring the voltage. Figure 4.1 gives an example of typical battery output data. The galvanostatic cycling gives a direct estimation of the resulting capacity of the electrodes as well the overpotential, which needs to be applied to charge the battery as compared to the discharge voltage. By monitoring the capacity over a number of cycles, the battery cyclability in terms of capacity retention is measured. These factors are often highly dependent on the magnitude of the applied current, which is conventionally described as the rate at which the battery is tested. Normally, a battery is tested at a series of different rates. The applied current is most commonly calculated based on the so-called C-rate, which is defined as the current required to discharge or charge the battery in one hour. However, for solid-state batteries – which are generally electrolyte-limited systems – it could be argued that the current density is a more relevant testing parameter, as it is analogous to the flux of Li^+ ions through the electrolyte.

The difference between the input and output capacity is moreover determined as the coulombic efficiency (CE):

$$CE = \frac{Q_{discharge}}{Q_{charge}} \tag{4.1}$$

Ideally, the $CE = 1$ but is often lower due to side reactions and different battery aging mechanisms, which make the battery gradually lose capacity. If the CE is <99%, the battery will have limited cyclability since the capacity deteriorates rather rapidly as a function of cycles unless there is an "unlimited" electrode present in the system, which can replenish the lost capacity – a typical example would be lithium metal in a lithium battery cell.

If battery testing data of SPE-based cells are compared to conventional cells using liquid electrolytes, a few things can generally be seen. First, the obtained capacity is often lower for SPE systems. This is due to a combination of larger overpotentials and inferior wettability, and a limited number of contact points between the

a)

b)

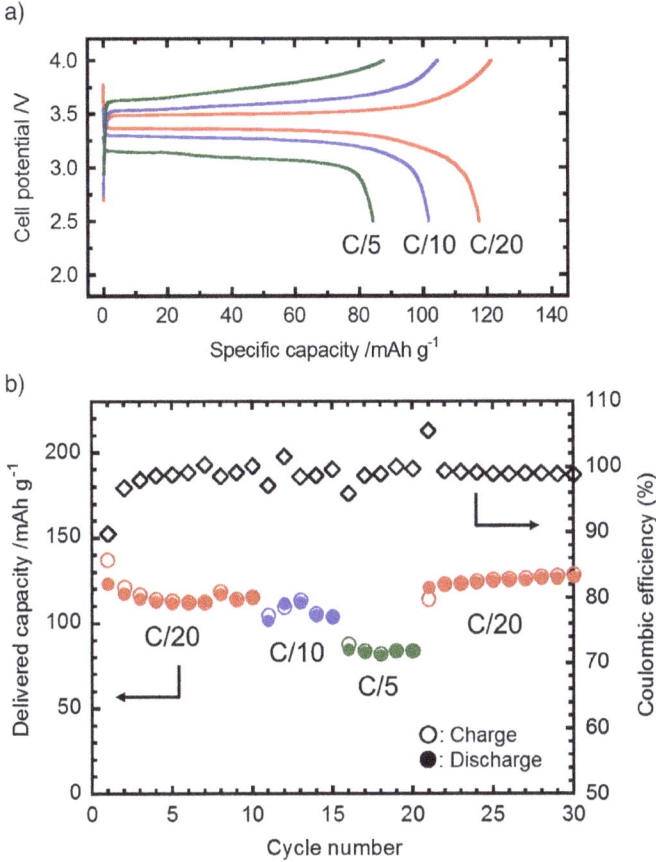

Fig. 4.1: SPE battery cycling of a LiFePO$_4$ | poly(ethylene carbonate):LiFSI | Li cell. (a) Voltage as a function of electrode capacity at different C-rates, with a clear overpotential visible as the voltage difference between charge and discharge. (b) Capacity retention tested for a series of C-rates. Reprinted from [6], Copyright 2016, with permission from Elsevier.

electrode and electrolytes. Therefore, often a gradual increase in capacity with cycling time can be observed [1, 2, 7], which is normally not the case for liquid-electrolyte counterparts. It is seen that as the cycling progresses, the polymer softens and diffuses into the porous electrode – ultimately filling all pores – whereafter the capacity ideally stabilizes around the theoretical capacity.

Second, a higher overpotential – that is, the difference between charge and discharge potentials, and thereby related to the energy efficiency – can be observed due to the higher resistivity of the SPE. If the voltage settings of the cycling are narrow, this can lead to a somewhat premature finish of the battery charging and thereby contributing to a reduced observed capacity. If the resistivity and overpotential increase during cycling, which is a common battery aging phenomenon, this is problematic to monitor and will appear as slowly degrading capacity, although the active material is intact in the electrode. Thus, predictability and uniformity of the SPE will lead to better possibilities for battery diagnostics. A third phenomenon also seen for SPE-based batteries, and likewise related to their limited conductivity, is their comparatively poor performance at higher rates. Often, the current rates applied are much lower than for conventional LIBs, unless the operating temperature is significantly higher. At the same time, it should be said that the tolerance for battery operation at elevated temperatures is much better for SPE-based systems, since the major aging mechanisms for LIBs are highly dependent on the liquid electrolyte applied [8, 9]. If SPE-based batteries are operated at a reasonably high temperature, that is, 70–80 °C, their performance in terms of low overpotential and good rate performance is quite attractive, without too much compromises in rapid battery aging. This is also why most commercial SPE battery systems have been targeting elevated temperature intervals [10, 11].

The CE when testing an SPE-based cell is often fluctuating, and not rarely approaching values above 100%, which seems anomalous at a first glance. This can to some degree be explained by the similarly fluctuating capacity observed due to the inferior electrode/electrolyte interfacial contacts, and which vary during the cycling [12]. Moreover, the fluctuations in CE can be caused by inhomogeneities in the electrodeposition of lithium, which leads to the deposition of dendritic lithium, and by formation of lithium deposits electrically and/or electrochemically isolated from the electrode (so-called dead lithium) [13, 14]. In addition, it is sometimes observed that instabilities occur during charge and which are seen as voltage noise, typically at high voltages. This can be attributed to side reactions due to poor electrochemical stability of the SPE at these voltages and/or poor mechanical stability of the SPE leading to formation of dendritic lithium [4, 15].

Abnormal behavior can sometimes be observed during SPE-based battery cycling. This can, at least for early stages of battery testing, be seen as a surprisingly good (!) battery performance. Not seldom is this seen as rate performance that does not comply with the bulk conductivity of the SPE – that is, the battery displays an unreasonably high capacity also at high current strengths for temperatures where the bulk Li^+ conductivity is rather low. This could be explained by solvent residues being incorporated in the SPE matrix when casting for the battery tests, while the casting for conductivity measurements is performed under different conditions. Generally, solvent residues can both plasticize the polymer and improve the surface wettability, and can render superior behavior during short-term testing. However, long-term performance, safety and predictability can suffer from the same cause. Furthermore, contaminations and residues can also lead to other battery problems, such as a poor CE due to side-reactions and an increased tendency for nucleation of dendritic lithium.

One specific problem for SPE-based battery cells is that corrosion of the aluminum current collector needs to be addressed differently than for conventional LIBs. In the latter, $LiPF_6$ is the salt normally employed. This salt undergoes a spontaneous reaction with Al, forming a passivating layer of AlF_3 on the current collector and protecting it from corrosion that otherwise constitutes a problem at high potentials. In SPEs, more chemically stable salts such as LiTFSI and LiFSI are normally employed, partly due to that they provide higher conductivity since they possess a low lattice enthalpy and good plasticizing ability, and partly due to that $LiPF_6$ has shown to be prone to side reactions in some SPEs, most notably those using PEO. This, in turn, might be due to the hygroscopic nature of PEO or the limited temperature stability of $LiPF_6$, which is less compatible with the operational temperature range of PEO-based SPEs. However, the C–F bond in TFSI and FSI is comparatively strong, and they therefore do not decompose to passivate the aluminum current collector [16]. This issue has been addressed for several different electrolyte systems [17]. On the one hand, the contact points between the current collector and the SPE are much more limited than for liquid electrolytes, which should cause less corrosion. Moreover, Al^{3+} ions are not soluble in, for example, PEO, and aluminum has shown to be resistant to corrosion below 3.8 V versus Li^+/Li even in the presence of LiTFSI for low current densities (0.05 mA cm^{-2}) [18]. On the other hand, potentiodynamic and potentiostatic polarization experiments used in corrosion cells do not accurately represent the real conditions of batteries. While corrosion of the aluminum current collector is not very significant for lithium metal batteries with PEO:LiTFSI when employing low-voltage cathodes [19], during overcharging with higher current densities or with high-voltage cathodes, aluminum is susceptible to pitting corrosion. Regions of high conductivity in the cathode or sites with high current densities lead to a breakdown of the protective film on aluminum [19, 20]. Introducing fluorinated polymers, such as perfluoropolyethers, that can interact with the TFSI anions [21] and decrease the amount of "free" TFSI anions can rectify these issues. There is also coated aluminum foil to use as current collectors, which can mitigate corrosion problems.

4.2 Compatibility with metal electrodes

As pointed out in Chapter 1, one of the primary reasons for employing SPEs is the potential realization of metal electrodes, which are otherwise problematic in combination with most liquid electrolytes. Metal electrodes will have a profound positive impact on the energy density of almost any battery system, which will make the batteries much more versatile. This is especially true for Li metal, which is also a metal electrode that is easy to use as compared to other high-energy-density battery metal electrodes (Na, K, Mg, etc.). Moreover, the infiltration problem that plagues the use of SPEs in porous electrodes is by comparison much more easily managed for these systems, where there is often a straightforward and energetically positive interaction between the metal electrode and the polymer host of the SPE. Metal–polymer bonds can easily be formed, which benefit surface adhesion and can potentially improve ion transport over the electrode–electrolyte interface [22].

The main challenge with the Li metal electrode is its tendency to form dendritic metal deposition during battery cycling. Long and needle-like metal structures of metallic lithium often appear growing into the electrolyte region due to uneven lithium deposition during the reduction process of battery charging. Generally, these result in a loss of the active amount of lithium in the cell and increased electrolyte consumption, but can also lead to safety hazards due to internal short circuits. In early modeling studies of lithium deposition [23], it was predicted that the formation of lithium dendrites is dependent on the shear modulus of the electrolyte material, and that a modulus of 7 GPa effectively blocks all opportunities for dendrite formation. While lower values of the shear modulus can still be applicable, and have shown effective in recent research [24], it is generally so that a more robust material can more effectively hinder dendrite formation. This means that polymer materials are more effective than liquid counterparts, but also that the mechanical properties of the SPE are vital for this process. As discussed in Chapter 3, it should also be kept in mind that there is a trade-off between mechanical rigidity and ion conduction for systems exhibiting coupled ion–polymer transport modes of conductivity. Furthermore, it is frequently argued that a high cationic transference number and a good wetting of the lithium metal will also contribute to mitigate lithium dendrite growth [25]. These parameters create a more uniform Li^+ ion flux close to the Li metal surface, thereby preventing inhomogeneous deposition.

Lithium nucleation is key for lithium dendrite growth, as Li^+ ions have a lower deposition interface energy on certain active sites. Regulating Li nucleation is thereby one way to control the dendrite growth. Here, the lithium metal and its surface play an important role. There is a difference in electrochemical behavior for different Li-metal foils – these are not entirely uniform. It is also so that it is the shear modulus at the very surface of Li that is relevant – this property is largely unknown, but might differ significantly from the bulk properties due to surface adhesion and local structural rearrangements. After nucleation, metallic lithium growth occurs and many different types of morphology may be formed that depend mainly on the applied current density and the mechanical properties of the SPE, but also on the salt concentration, tip radius of the protrusion, temperature, pressure, solid electrolyte interphase and the ion transport properties of the electrolyte [26]. At low current densities (below the limiting current) lithium grows at the base or roots of the dendritic protrusion forming a "mossy" morphology. At higher currents (above the limiting current), at the onset of electrolyte diffusion limitation, lithium instead deposits at the tips of the dendritic protrusion forming "dendritic" structures. The latter are harder to control and can lead to short circuiting [27, 28]. Thereby, in order to prevent dendrite growth, the applied current should be below the limiting current of the SPE, a mechanical stress field should be applied and the SPE should therefore have a high shear modulus and a high yield strength. This latter property is often disregarded [27].

Dendrite formation on lithium metal can be tested either in battery half-cells (*i.e.*, employing a Li-metal anode) or more systematically in symmetrical Li | SPE | Li cells. If dendrite formation is taking place, this then appears as voltage noise in the voltage profile of battery half-cells [4] or as a sharp drop in the magnitude of the voltage in symmetrical lithium cells as shown in Fig. 4.2 [29–33]. Several approaches have been taken to delay or suppress dendrite growth. Increasing the modulus of the SPE by for example using SEO – a copolymer of polystyrene and PEO – instead of PEO has shown to delay the dendrite short circuit from a few days to a few months [29]. Increasing the thickness of the SPE increases the Li dendrite path and higher force would be required to penetrate the material [4], but this also renders a higher resistivity. Another strategy to tackle the dendrite growth issue in PEO is to include it in a semi-interpenetrating network. In such an SPE, higher compressive strain and stress are achieved and the material demonstrates a better ability to withstand volume changes and external forces [34]. In contrast, another example of a cross-linked polyethylene–PEO SPE with low modulus has been reported to have excellent resistance to dendrite growth, suggesting that high-modulus SPEs are not always required [35].

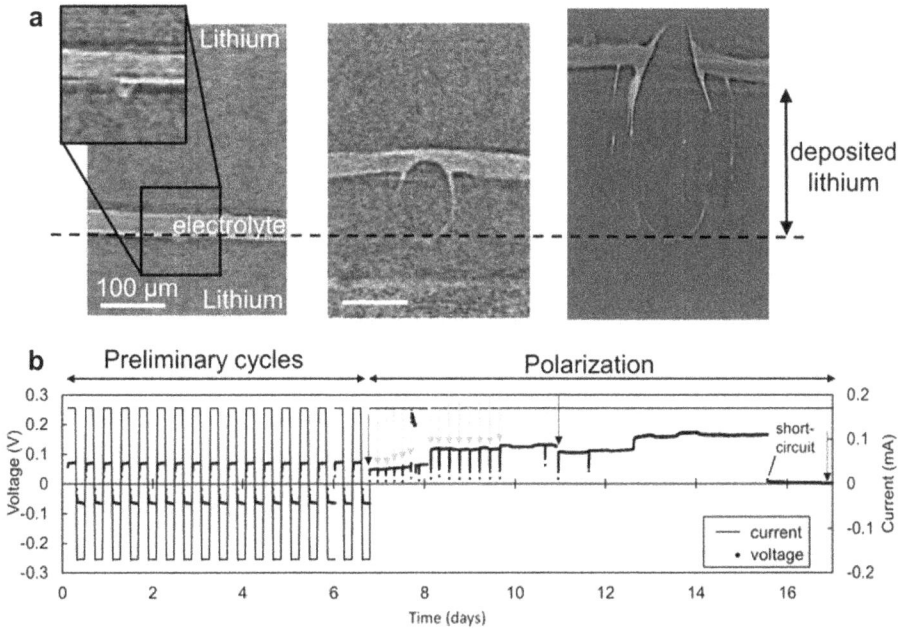

Fig. 4.2: (a) Sequence of X-ray microtomography images showing the growth of a lithium globule that eventually punctures the polymer electrolyte membrane and short circuits the cell. (b) Cycling profile of a symmetric Li | SPE | Li cell during preliminary cycling, polarization and afterwards short circuiting. The images in (a) correspond to the black arrows in (b). Reprinted from [33] under CC-BY 4.0.

As seen from these aforementioned examples, it is sometimes stated that block copolymers and/or cross-linking will lead to dendrite suppression, due to the increase in modulus – often with orders of magnitude. Theoretically, this is somewhat counterintuitive, since the ionic current that controls Li nucleation will travel through regions, which microscopically behave very similarly to a homogeneous polymeric system. Since there are still soft polymer regions present where the ionic current preferably goes, this is likely where the dendrites will grow. On the other hand, the mesoscale structures formed due to these more advanced polymer architectures might also play a role in the possibility for sharp needle-like structures to form and penetrate the SPE membrane, at the same time as there are other things to gain from a more mechanically robust electrolyte, not least its ability to act as separator [36].

Surface decomposition, SEI formation and stability of the electrode–electrolyte interfaces are key factors that determine the battery performance. However, they have so far not really been extensively studied for SPEs, mainly due to the arduousness in separating the battery components in solid-state cells, especially after cycling. Thus, this makes it harder for polymer-based solid-state batteries to use conventional interfacial analytical techniques that have been developed for liquid-based batteries, in particular X-ray photoelectron spectroscopy (XPS) [37]. The most widely used technique is instead electrochemical impedance spectroscopy, which can provide information about resistance changes at the different interfaces in situ in the cell [38]. While scanning electron microscopy (SEM) has been broadly employed to study dendrite growth, it has also been used for investigating interfacial degradation and active material dissolution into the electrolyte [39, 40]. Surface analysis of the decomposition layer formed between the SPE and the electrode has also been carried out in a few studies with XPS, which has provided useful information about the chemical composition of the interfacial layers with a high degree of elemental sensitivity. The main SEI components reported with SPEs are Li_2CO_3, Li_2O and LiF [41].

The many issues discussed above occurring in lithium metal batteries are also present for other metal electrodes, for example, sodium metal batteries, where the growth of sodium dendrites leads to short circuits [42] and the highly reactive sodium metal results in instabilities at the polymer electrolyte–sodium interface [43, 44]. Due to the higher chemical reactivity of sodium than lithium, these challenges are also greater than in lithium batteries – and likewise for many other batteries employing metal electrodes. Despite all its problems, metallic lithium is one of the least problematic electrodes to use with SPEs.

4.3 Compatibility with porous electrodes

Conventional liquid-electrolyte lithium-ion batteries benefit from a good contact between electrolyte and electrodes thanks to the ability of the liquid to infiltrate in the porous electrode. Unfortunately, that is not the case for solid-state electrolytes where the electrochemical reactions occur through a solid–solid interface between the electrode and the electrolyte, and this becomes increasingly difficult when going from a metal electrode to one made up of several particles forming a porous matrix. Formation of intimate contacts at the electrode particle/electrolyte interface is key to enhance the charge-transfer reactions and thus improve the electrochemical performance of the solid-state batteries. Although polymer electrolytes have a better interfacial contact with the electrode compared to ceramic or inorganic electrolytes, the resulting resistance is often still quite high. In this regard, a lot of effort has been made to improve the interfacial contact between the polymer electrolyte and the electrode, as well as to solve the infiltration problem.

One solution to improve the wettability and interfacial contact in porous electrodes while maintaining a simple implementation method of the SPE is to incorporate a small amount of oligomers between both components. This has been shown to improve the initial capacity of the battery [1]. Alternatively, casting a solution of the SPE directly onto the prefabricated porous electrode is another way to improve the interfacial contact with the active material as the SPE material can fill the pores of the cathode when it is in solution (Fig. 4.3) [1, 45]. With this method, the solvent used to dissolve the polymer and salt should not dissolve the binder of the cathode, since this will consequently lead to it losing its integrity. Moreover, solvent can this way be trapped in the porous electrode, and later react chemically or electrochemically during battery operation.

Another advantage of SPEs is that the polymer material can often be used as binder in the electrode and replace the conventional inactive binders. This can improve the ionic conductivity in the electrode, while a good chemical compatibility between the binder on the electrode particles and the SPE material can also generate beneficial contacts with the active material. In order for the SPE to be used as binder, it should be mechanically stable at the operating temperature and provide good binding properties. This approach has been used with PEO-based SPEs [46], and the effects of incorporating SPEs as binders and comparing it with conventional binders has also been systematically investigated for carbonyl-based SPEs [7, 47]. These studies have shown that incorporating the polymer host material as binder in the cathode formulation leads to a lower resistance and polarization during cycling compared to the conventionally used PVdF (poly(vinylidene fluoride)) binder. Therefore, it is suggested that combining both the above approaches – including SPE as binder as well as using infiltration casting on top of the porous cathode – will further improve the mass transport within the inner parts of the cathode and thereby decreasing polarization and resistance [47]. Having a good contact between the components is not the only key factor for a better performance, but a good chemical and mechanical compatibility is also required. In this regard, it has been suggested that a high-modulus SPE binder prevents the formation of a good electrolyte–electrode interface. In contrast, a softer and low-molecular-weight SPE as binder forms a more compatible interface between the binder and the electrode, as well as with the bulk electrolyte, facilitating ion transfer over the phase boundary [7]. Similarly, there can in principle be specific interactions between the carbon additive and the SPE, which can provide different properties that might affect SPE-active material interaction, or the ability for the carbon additive

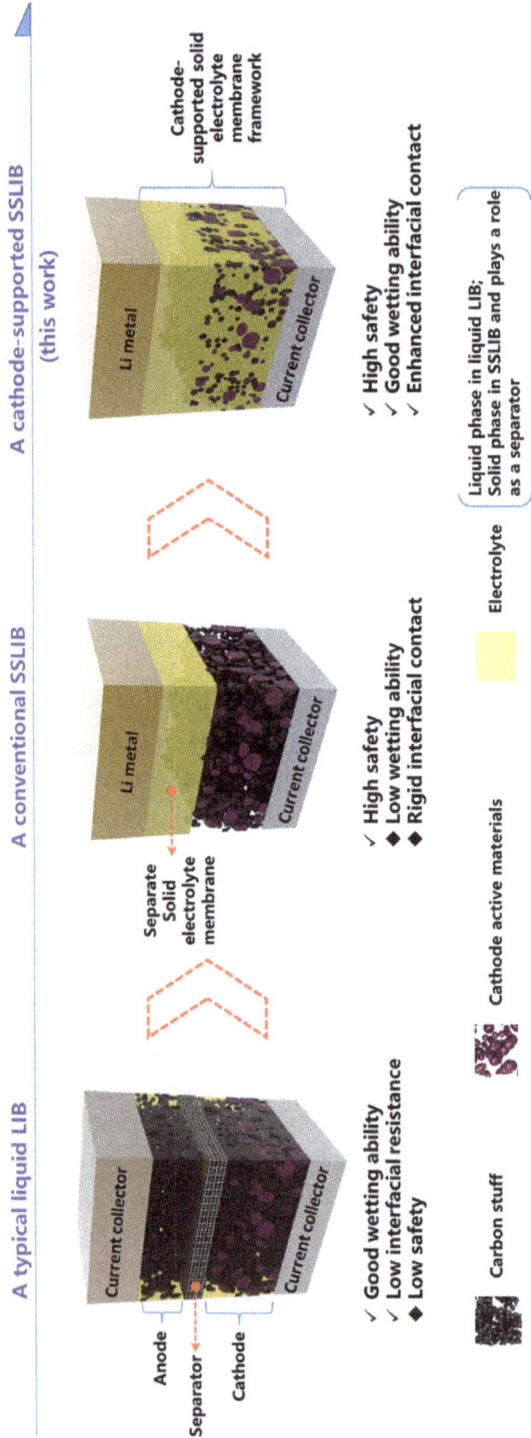

Fig. 4.3: Schematic of the good wetting ability of liquid electrolytes (left), poor wetting ability of solid-state electrolytes (center) and improved wettability when casting the SPE directly onto the cathode electrode. Republished with permission of The Royal Society of Chemistry from [45]; permission conveyed through Copyright Clearance Center, Inc.

to form a good electronic wiring. There should clearly be room for tailoring and optimizing the interactions between the SPE polymer and all components in the electrode: active material, binder and carbon additive. So far, this is largely unexplored territory in the scientific literature. Overall, however, it can be concluded that using a binder that is soft and chemically similar to the SPE, and solvent-casting the SPE onto the cathode, leads to better compatibility and an improved electrolyte–electrode interface lowering the resistance and polarization, thereby improving the battery performance.

Besides the importance of the compatibility with metal anodes, the ability of solid-state electrolytes to be paired with different cathode materials is just as critical, but generally less explored. The most commonly reported cathode material used in polymer-based solid-state batteries is LiFePO$_4$ (LFP). LFP operates at around 3.5 V versus Li$^+$/Li, which is clearly within the electrochemical stability window of most reported SPEs. However, it is very common that also wider ESWs are reported as electrolyte properties, but more rarely tested in true battery devices. In order to access the high-voltage cathodes used nowadays in LIBs such as LiNi$_x$Co$_y$Al$_z$O$_2$ (NCA) (4.3 V vs Li$^+$/Li), LiNi$_x$Mn$_y$Co$_z$O$_2$ (NMC) (4.5 V vs Li$^+$/Li) and even higher-voltage materials such as LiNi$_{0.5}$Mn$_{1.5}$O$_4$ (LNMO) (4.9 V vs Li$^+$/Li), SPEs that are electrochemically stable at high voltages are indeed required. In addition, there are other battery technologies beyond the conventional LIBs with higher capacities that could benefit from a solid-state electrolyte, such as lithium–sulfur batteries, lithium–oxygen batteries and organic batteries, but that will pose additional requirements and challenges for the SPE.

PEO, the classic solid polymer electrolyte host material, has repeatedly shown a limited electrochemical stability at high voltages. This constitutes a major setback for its implementation with high-voltage cathode materials, and thereby for high-energy-density batteries. In fact, it has often shown to be difficult to cycle PEO-based SPEs with any other common LIB cathode than LFP. On the other hand, there are other families of SPEs that are more stable at higher voltages, such as carbonyl-based SPEs [48, 49], and nitrile-based SPEs [50]. For example, a polymer combining ether and carbonate groups in the backbone was reported to provide good battery performance at 25 °C with LiFePO$_4$ (3.5 V vs Li$^+$/Li) as well as with the other high-voltage cathode LiFe$_{0.2}$Mn$_{0.8}$PO$_4$ (4.1 V vs Li$^+$/Li) [48]. Fluorinated polycarbonates have also reported improved electrochemical performance versus LiCoO$_2$ compared to a PEO-based SPE, which can be attributed to the higher oxidative stability of the carbonate, the fluorinated groups and the cyano end-groups of the polymer [2].

Often, while trying to improve the electrochemical stability toward high voltages on the cathode side, this results in materials where the stability is detrimental at the low voltages, which are present at the anode side. To mitigate this issue, it has been suggested in recent years that a *double-layer* polymer electrolyte could be employed. Here, a layer of an SPE stable at low voltages placed facing the lithium metal is combined with a layer of another SPE stable at higher voltages, placed toward the cathode. Solid electrolytes, which do not mix like most liquid counterparts, could therefore be used to locally tailor the electrolyte system at the two different electrodes. This strategy has been studied with a PEO-based SPE in contact with lithium metal and a poly(*N*-methyl-malonic amide) in contact with the cathode, more specifically $LiCoO_2$ [51]. It has been claimed that such a system provides stable interfaces with no visible reduction of Li^+ conductivity across the additional polymer/polymer interface introduced by this approach.

Nevertheless, it is striking that the largest problems observed for the SPEs in terms of chemical and electrochemical side reactions seem to appear on the cathode rather than the usually more reactive anode – this is in stark contrast to liquid electrolytes. One additional challenge related to the cathode side is its volume expansion/contraction during battery cycling. Although the volume change is rather low compared to the anodes, it is sufficient to damage the electrode–electrolyte interface. This loss of contact could limit the accessible capacity or power that would require additional mechanical force – employed by for example external pressure – to maintain contact during operation. However, such high pressures could also limit the material's freedom to expand during cycling. In addition, compatibility of the SPE in contact with the cathode active material is required, particularly at high temperatures and high voltages, to ensure good battery performance [52]. Also these aspects call for more profound future research.

SPEs, which are inherently more chemically stable and better prevent diffusivity of reaction products from the electrodes than liquid electrolytes, could be envisioned to have superior functionality with a range of reactive and/or less stable electrode materials. One such application for SPEs could be in organic batteries. Replacing the inorganic materials in the cathode with organic molecules is often highlighted as potentially rendering lower-cost batteries that are more environmentally friendly. However, one of the main challenges for this type of batteries is the continuous dissolution of the active material into the liquid electrolyte. Therefore, the use of solid polymer electrolytes could mitigate or eliminate this issue, making this type of batteries more competitive. In addition, since the use of SPEs can permit the use of lithium metal and consequently the implementation of non-lithiated organic molecules, this opens up a larger choice of available electrode materials. Initially, gel polymer electrolytes were studied using PVdF and PEO as gelling agents for this purpose [53, 54]. However, the presence of large amounts of liquid did not solve the dissolution issue of organic electrodes. Instead, replacing the liquid fraction with SiO_2 particles resulted in improved performance using a pillar[5]quinone cathode and poly(methacrylate)/PEO:$LiClO_4$–SiO_2 electrolyte [55]. Another example of an organic battery with a solid polymer electrolyte uses a tetramethoxy-*p*-benzoquinone (TMQ) as active material and PEO:LiTFSI as the SPE. This battery was able to provide higher reversible capacity and better capacity retention at 100 °C than the reference cell with liquid electrolyte at 20 °C. Even though material dissolution and diffusion of the organic molecule through the SPE were observed, the surface of the lithium electrode appeared stable, and no self-discharge-inducing redox shuttle processes or major decomposition products were detected [56].

Another type of problematic but promising battery technology is lithium–sulfur batteries that combine high-capacity electrodes (3,800 mAh g^{-1} of lithium and 1,675 mAh g^{-1} of sulfur) and can thereby theoretically yield a high energy density of 2,600 Wh kg^{-1} for an operating voltage of 2 V. The main challenges of this technology are the lithium dendrite formation and the soluble intermediate products (polysulfides) that migrate from the cathode during battery operation, and react with the lithium metal anode [57]. One unified way to overcome these two issues would be the use of solid-state electrolytes that can potentially block both dendrite growth and the polysulfide shuttle. As far back as in the year 2000, different kinds of PEO-based SPEs were employed in Li–S batteries. Despite good initial discharge capacity, a rapid capacity fade (within the first 20 cycles) limited the useful lifetime of these cells [58]. This was shown to be due to that the cathode's structure collapsed during the first cycle. Furthermore, despite the use of a solid polymer electrolyte, the formed polysulfides swelled the SPE, diffused through it and reacted with the lithium metal, forming a thick layer of Li$_2$S at the Li/electrolyte interface. These processes led to a continuous loss of active material and consequently a fast capacity decay. In more recent work, the importance of maintaining the initial structure of the electrode to get a good cyclability has been highlighted. This can be achieved by enhancing the mechanical strength of the cathode and completely avoiding the polysulfide diffusion [40].

An even higher theoretical energy density than for lithium–sulfur can be obtained with a lithium–oxygen battery (3,500 Wh kg^{-1}). The challenges of this technology are the decomposition of organic liquid electrolytes, the blockage of air diffusion in the porous cathode by the insoluble discharge products, and lithium dendrites that lead to large polarization, capacity decay and safety concerns [57]. Developing solid-state lithium–oxygen batteries could potentially solve some of these issues. Most of the reported results use composite or inorganic electrolytes [59, 60], but also truly solid polymer electrolyte lithium–oxygen batteries have been investigated, for example, using a PEO-based SPE at 80 °C. The corresponding battery provided improved electrochemical performance, higher discharge voltage and lower charge voltage, compared to the analogous glyme-based liquid electrolyte. Although the capacity was slightly lower, the results demonstrate the feasibility to use a solid polymer electrolyte in lithium–oxygen batteries [61].

Yet one interesting approach where SPEs can provide a promising alternative is for 3D-microbatteries (Fig. 4.4), which can be rendered based on 3D-printing, lithography, electrochemical deposition, carbon foams, etc. [62]. These devices, often mm^3 in size, try to optimize power and energy density simultaneously by utilizing advanced battery architectures where the electrodes are interdigitated with each other. The electrolyte needs to be ultrathin yet robust, and be able to coat the electrode structures uniformly, conformally and pinhole-free. While liquid electrolytes cannot separate the electrodes, ceramic electrode materials are difficult to deposit into these highly porous structures. Promising strategies for SPEs have been to use monomers that can self-assemble on the complex electrode surfaces, and then cross-link these polymers to form a robust coating [63]. The polymers can also be functionalized with surfactant groups, such as –OH, which can adhere or graft to the electrode surfaces, resulting in stable yet thin layers that can function electrochemically [64].

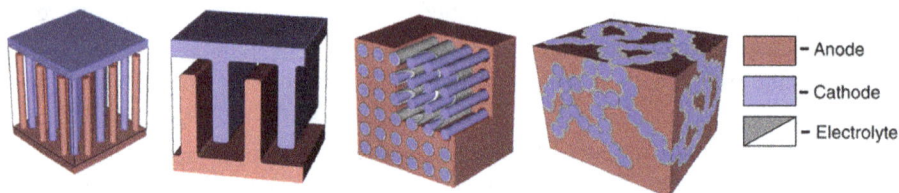

Fig. 4.4: Different conceptual 3D-microbattery architectures. Adapted from [65], Copyright 2011, with permission from Elsevier.

4.4 Processing and use of large-scale SPE-based batteries

The manufacturing process of polymer-based solid-state batteries is very similar to traditional LIB manufacturing, but with some important differences [66]:
- There is no need for anode coating as these batteries more or less exclusively use lithium metal as anode.
- There is no need for electrolyte filling before sealing of the cell, as there is no liquid electrolyte.
- Instead, an additional second coating step is required to implement the solid polymer electrolyte.
- Formation cycles are not required, which are usually time-consuming and costly.

Processing of solid polymer electrolytes in batteries involves three main steps: (i) mixing the components, (ii) shaping or coating them onto an electrode or a substrate and (iii) a compaction step to ensure sufficient mechanical contact, low interfacial resistance and high density. The type of SPE and its properties will determine the specific process to be used, where generally the most important step is the second (ii) which involves shaping or coating the polymer. In addition, another important aspect of the SPE manufacturing process is the local environment and atmospheric conditions. While conventional LIBs and some inorganic solid-state electrolytes are prone to generate toxic compounds such as HF and H_2S, respectively, SPE-based batteries in contrast often do not require any special risk mitigation strategies. However, similar to other battery manufacturing, the process has to be carried out in dry or inert atmosphere, as moisture residues will have a negative impact on battery performance. This is particularly important for common SPEs that employ LiTFSI salt and PEO, which are both hygroscopic.

The manufacturing processes for large-scale SPE-based batteries are either based on solvent casting or extrusion. The traditional solvent casting methods used on a lab scale can generally be upscaled straightforwardly. Such processes can be implemented in large-scale manufacturing processes with tape casting, slot die or doctor blade casting, or by screen printing onto one of the electrodes or another substrate. In these processes, also referred to as wet chemical processing, the polymer and the salt are dissolved in a solvent, where viscosity and rheological properties are key parameters to be considered. Such a solution or slurry is then applied directly onto one of the electrodes or onto a different substrate. In the former case, the SPE will be infiltrated into the cathode's pores, and it is important to consider that the solvent used should not dissolve the cathode binder [67]. In the latter case, one additional step is required to place the SPE on top of the electrode. After coating, the solvents are evaporated in a drying step. Finally, a compaction process is applied by calendaring, pressing or heat treatment. The main drawback of this process is the use of a solvent that requires additional drying for evaporation, and which can also bring associated environmental concerns. Figure 4.5 displays an example of this process.

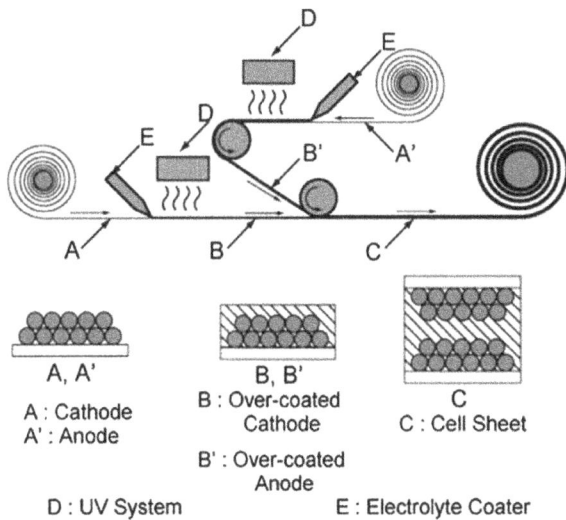

Fig. 4.5: Schematic of a large-scale manufacturing process for solid-state batteries using wet chemical processing. Reprinted from [68], Copyright 2008, with permission from Elsevier.

Another well-established manufacturing process for polymers is extrusion, and this can also be applied for a solid polymer electrolyte. This is a dry process, solvent-free, rapid and cost-effective. Thus, it has already been applied for large-scale polymer-based solid-state batteries [69, 70]. In this process, depicted in Fig. 4.6, the polymer and salt are fed into a high-temperature extruder, where they are melted, mixed and finally extruded through a slit die and discharged onto a substrate sheet or electrode. A compaction step is implemented afterward through cylinder rollers. Also this process possesses several challenges. Some polymers with low melting point (such as PEO) are not stable under conventional extrusion conditions, which employ high temperature and high shear rate. Another challenge is to further improve the efficiency of the manufacturing process by simultaneously extruding multiple layers of cathode and polymer electrolyte [70].

The perhaps most well-known example of a commercial polymer-based solid-state battery manufacturer is Bolloré. The company started working on their technology in the 1990s, with industrialization in 2001 through the creation of the subsidiary Batscap. The cathode contains the active material (originally VO_x), conductive additive and an ionically conductive polymer electrolyte material composed of PEO and salt, and lithium metal as the anode. Introducing an ionically conductive polymer already as a binder is done to ensure good contact with the active material and to enhance the compatibility with the polymer electrolyte layer. The cathode mixture is extruded onto the Al-based current collector. At the same time, polymer and salt forming the bulk electrolyte are also extruded and deposited onto the cathode electrode (Fig. 4.6). Already in 1997, Bolloré reported promising results with 4 Ah and 40 Ah cells [69]. Nowadays, the cathode is based on LFP instead of VO_x. LFP is a nonpolluting material that is cobalt- and nickel-free, thus avoiding socially and environmentally problematic materials. All battery components can be recycled. The battery technology is used in electric vehicles, buses and stationary storage through the brand Blue Systems. The Bluecar's battery provides 30 kWh, the size is 300 L and 300 kg and the internal operating temperature is between 60 and 80 °C, but it can operate with external temperatures of −20 to 160 °C because of its rather low sensitivity to external temperature variations [72].

A lot of small and medium-sized enterprises are moving into SPE-based battery production today, while also larger companies are forecasting production. The development in the commercial sector is currently rapid, and it is difficult to make a robust market outlook within the context of this book. It is interesting to note, however, that the targeted systems span several different types of polymer chemistries and polymer architectures, and that several polymer–ceramic composites have been highlighted for commercial development of solid-state batteries.

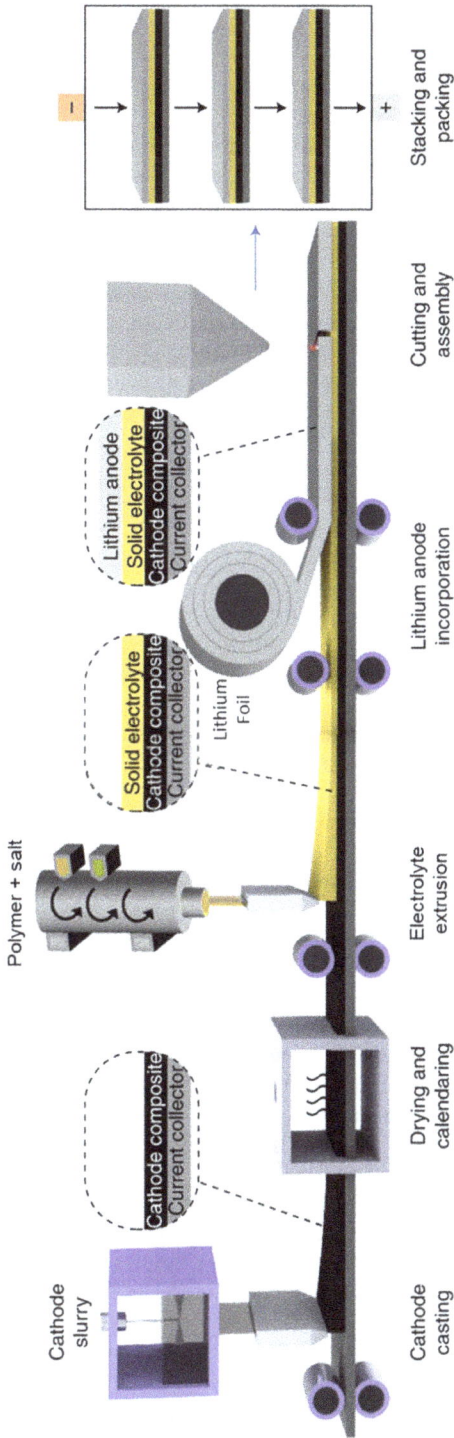

Fig. 4.6: Schematic of the process chain for solid electrolyte fabrication using extrusion. Adapted from [5], Copyright 2018, with permission from Elsevier and by permission from Springer Nature Customer Service Centre GmbH [71], 2020.

References

[1]   Sun B, Mindemark J, Edström K, Brandell D. Realization of high performance polycarbonate-based Li polymer batteries. Electrochem Commun. 2015;52:71–4.

[2]   Wang Q, Liu X, Cui Z, Shangguan X, Zhang H, Zhang J, et al. A fluorinated polycarbonate based all solid state polymer electrolyte for lithium metal batteries. Electrochim Acta. 2020;337:135843.

[3]   Gray FM, MacCallum JR, Vincent CA. Poly(ethylene oxide) – LiCF3SO3 – polystyrene electrolyte systems. Solid State Ionics. 1986;18–19:282–6.

[4]   Homann G, Stolz L, Nair J, Laskovic IC, Winter M, Kasnatscheew J. Poly(ethylene oxide)-based electrolyte for solid-state-lithium-batteries with high voltage positive electrodes: Evaluating the role of electrolyte oxidation in rapid cell failure. Sci Rep. 2020;10:4390.

[5]   Schnell J, Günther T, Knoche T, Vieider C, Köhler L, Just A, et al. All-solid-state lithium-ion and lithium metal batteries – paving the way to large-scale production. J Power Sources. 2018;382:160–75.

[6]   Kimura K, Yajima M, Tominaga Y. A highly-concentrated poly(ethylene carbonate)-based electrolyte for all-solid-state Li battery working at room temperature. Electrochem Commun. 2016;66:46–8.

[7]   Bergfelt A, Hernández G, Mogensen R, Lacey MJ, Mindemark J, Brandell D, et al. Mechanically robust yet highly conductive diblock copolymer solid polymer electrolyte for ambient temperature battery applications. ACS Appl Polym Mater. 2020;2:939–48.

[8]   Palacín MR, De Guibert A. Why do batteries fail? Science. 2016;351:1253292.

[9]   Birkl CR, Roberts MR, McTurk E, Bruce PG, Howey DA. Degradation diagnostics for lithium ion cells. J Power Sources. 2017;341:373–86.

[10]   Genieser R, Loveridge M, Bhagat R. Practical high temperature (80 °C) storage study of industrially manufactured Li-ion batteries with varying electrolytes. J Power Sources. 2018;386:85–95.

[11]   https://www.blue-solutions.com/en/blue-solutions/technology/batteries-lmp/

[12]   Yang H, Zhang Y, Tennenbaum MJ, Althouse Z, Ma Y, He Y, et al. Polypropylene carbonate-based adaptive buffer layer for stable interfaces of solid polymer lithium metal batteries. ACS Appl Mater Interfaces. 2019;11:27906–12.

[13]   Chen K-H, Wood KN, Kazyak E, LePage WS, Davis AL, Sanchez AJ, et al. Dead lithium: Mass transport effects on voltage, capacity, and failure of lithium metal anodes. J Mater Chem A. 2017;5:11671–81.

[14]   Choudhury S, Tu Z, Nijamudheen A, Zachman MJ, Stalin S, Deng Y, et al. Stabilizing polymer electrolytes in high-voltage lithium batteries. Nat Commun. 2019;10:3091.

[15]   Zhao Q, Chen P, Li S, Liu X, Archer LA. Solid-state polymer electrolytes stabilized by task-specific salt additives. J Mater Chem A. 2019;7:7823–30.

[16]   Kanamura K, Umegaki T, Shiraishi S, Ohashi M, Takehara Z-I. Electrochemical behavior of Al current collector of rechargeable lithium batteries in propylene carbonate with LiCF3SO3, Li(CF3SO2)2N, or Li(C4F9SO2)(CF3SO2)N. J Electrochem Soc. 2002;149:A185.

[17]   Matsumoto K, Inoue K, Nakahara K, Yuge R, Noguchi T, Utsugi K. Suppression of aluminum corrosion by using high concentration LiTFSI electrolyte. J Power Sources. 2013;231:234–8.

[18]   Li Q, Imanishi N, Takeda Y, Hirano A, Yamamoto O. PEO-based composite lithium polymer electrolyte, PEO-BaTiO3-Li(C2F5SO2)2N. Ionics. 2002;8:79–84.

[19]   Chen Y, Devine TM, Evans JW, Monteiro OR, Brown IG. Examination of the corrosion behavior of aluminum current collectors in lithium/polymer batteries. J Electrochem Soc. 1999;146:1310–17.

[20] Wetjen M, Kim G-T, Joost M, Appetecchi GB, Winter M, Passerini S. Thermal and electrochemical properties of PEO-LiTFSI-Pyr14TFSI-based composite cathodes, incorporating 4 V-class cathode active materials. J Power Sources. 2014;246:846–57.

[21] Cong L, Liu J, Armand M, Mauger A, Julien CM, Xie H, et al. Role of perfluoropolyether-based electrolytes in lithium metal batteries: Implication for suppressed Al current collector corrosion and the stability of Li metal/electrolytes interfaces. J Power Sources. 2018;380: 115–25.

[22] Lopez J, Pei A, Oh JY, Wang G-JN, Cui Y, Bao Z. Effects of polymer coatings on electrodeposited lithium metal. J Am Chem Soc. 2018;140:11735–44.

[23] Monroe C, Newman J. The impact of elastic deformation on deposition kinetics at lithium/ polymer interfaces. J Electrochem Soc. 2005;152:A396.

[24] Nair JR, Shaji I, Ehteshami N, Thum A, Diddens D, Heuer A, et al. Solid polymer electrolytes for lithium metal battery via thermally induced Cationic Ring-Opening Polymerization (CROP) with an insight into the reaction mechanism. Chem Mater. 2019;31:3118–33.

[25] Zhou W, Wang S, Li Y, Xin S, Manthiram A, Goodenough JB. Plating a dendrite-free lithium anode with a polymer/ceramic/polymer sandwich electrolyte. J Am Chem Soc. 2016;138: 9385–8.

[26] Frenck L, Sethi GK, Maslyn JA, Balsara NP. Factors That control the formation of dendrites and other morphologies on lithium metal anodes. Front Energy Res. 2019;7.

[27] Barai P, Higa K, Srinivasan V. Lithium dendrite growth mechanisms in polymer electrolytes and prevention strategies. Phys Chem Chem Phys. 2017;19:20493–505.

[28] Bai P, Li J, Brushett FR, Bazant MZ. Transition of lithium growth mechanisms in liquid electrolytes. Energy Environ Sci. 2016;9:3221–9.

[29] Stone GM, Mullin SA, Teran AA, Hallinan DT, Minor AM, Hexemer A, et al. Resolution of the modulus versus adhesion dilemma in solid polymer electrolytes for rechargeable lithium metal batteries. J Electrochem Soc. 2012;159:A222–A7.

[30] Sun F, Moroni R, Dong K, Markötter H, Zhou D, Hilger A, et al. Study of the mechanisms of internal short circuit in a Li/Li Cell by synchrotron X-ray phase contrast tomography. ACS Energy Lett. 2017;2:94–104.

[31] Rosso M, Brissot C, Teyssot A, Dollé M, Sannier L, Tarascon J-M, et al. Dendrite short-circuit and fuse effect on Li/polymer/Li cells. Electrochim Acta. 2006;51:5334–40.

[32] Brissot C, Rosso M, Chazalviel JN, Lascaud S. Dendritic growth mechanisms in lithium/ polymer cells. J Power Sources. 1999;81–82:925–9.

[33] Harry KJ, Liao X, Parkinson DY, Minor AM, Balsara NP. Electrochemical deposition and stripping behavior of lithium metal across a rigid block copolymer electrolyte membrane. J Electrochem Soc. 2015;162:A2699–A706.

[34] Homann G, Stolz L, Winter M, Kasnatscheew J. Elimination of "voltage noise" of poly (ethylene oxide)-based solid electrolytes in high-voltage lithium batteries: Linear versus network polymers. iScience. 2020;23:101225.

[35] Khurana R, Schaefer JL, Archer LA, Coates GW. Suppression of lithium dendrite growth using cross-linked polyethylene/poly(ethylene oxide) electrolytes: A new approach for practical lithium-metal polymer batteries. J Am Chem Soc. 2014;136:7395–402.

[36] Young W-S, Kuan W-F, Epps I, Thomas H. Block copolymer electrolytes for rechargeable lithium batteries. J Polym Sci Part B: Polym Phys. 2014;52:1–16.

[37] Sångeland C, Mindemark J, Younesi R, Brandell D. Probing the interfacial chemistry of solid-state lithium batteries. Solid State Ionics. 2019;343:115068.

[38] Bouchet R, Lascaud S, Rosso M. An EIS study of the anode Li/PEO-LiTFSI of a Li polymer battery. J Electrochem Soc. 2003;150:A1385.

[39] Hovington P, Lagacé M, Guerfi A, Bouchard P, Mauger A, Julien CM, et al. New lithium metal polymer solid state battery for an ultrahigh energy: Nano C-LiFePO4 versus Nano Li1.2V3O8. Nano Lett. 2015;15:2671–8.

[40] Lécuyer M, Gaubicher J, Deschamps M, Lestriez B, Brousse T, Guyomard D. Structural changes of a Li/S rechargeable cell in lithium metal polymer technology. J Power Sources. 2013;241: 249–54.

[41] Xu C, Sun B, Gustafsson T, Edström K, Brandell D, Hahlin M. Interface layer formation in solid polymer electrolyte lithium batteries: An XPS study. J Mater Chem A. 2014;2:7256–64.

[42] Qiao L, Judez X, Rojo T, Armand M, Zhang H. Review – polymer electrolytes for sodium batteries. J Electrochem Soc. 2020;167:070534.

[43] Sångeland C, Mogensen R, Brandell D, Mindemark J. Stable cycling of sodium metal all-solid-state batteries with polycarbonate-based polymer electrolytes. ACS Appl Polym Mater. 2019;1:825–32.

[44] Zhao C, Liu L, Lu Y, Wagemaker M, Chen L, Hu Y-S. Revealing an interconnected interfacial layer in solid-state polymer sodium batteries. Angew Chem Int Ed. 2019;58:17026–32.

[45] Chen X, He W, Ding L-X, Wang S, Wang H. Enhancing interfacial contact in all solid state batteries with a cathode-supported solid electrolyte membrane framework. Energy Environ Sci. 2019;12:938–44.

[46] Kobayashi Y, Mita Y, Seki S, Ohno Y, Miyashiro H, Terada N. Comparative study of lithium secondary batteries using nonvolatile safety electrolytes. J Electrochem Soc. 2007;154:A677.

[47] Bergfelt A, Lacey MJ, Hedman J, Sångeland C, Brandell D, Bowden T. ε-Caprolactone-based solid polymer electrolytes for lithium-ion batteries: Synthesis, electrochemical characterization and mechanical stabilization by block copolymerization. RSC Adv. 2018; 8:16716–25.

[48] He W, Cui Z, Liu X, Cui Y, Chai J, Zhou X, et al. Carbonate-linked poly(ethylene oxide) polymer electrolytes towards high performance solid state lithium batteries. Electrochim Acta. 2017;225:151–9.

[49] Chai J, Liu Z, Ma J, Wang J, Liu X, Liu H, et al. In situ generation of poly (vinylene carbonate) based solid electrolyte with interfacial stability for LiCoO2 lithium batteries. Adv Sci. 2017;4:1600377.

[50] Hu P, Chai J, Duan Y, Liu Z, Cui G, Chen L. Progress in nitrile-based polymer electrolytes for high performance lithium batteries. J Mater Chem A. 2016;4:10070–83.

[51] Zhou W, Wang Z, Pu Y, Li Y, Xin S, Li X, et al. Double-layer polymer electrolyte for high-voltage all-solid-state rechargeable batteries. Adv Mater. 2019;31:1805574.

[52] Kerman K, Luntz A, Viswanathan V, Chiang Y-M, Chen Z. Review – practical challenges hindering the development of solid State Li ion batteries. J Electrochem Soc. 2017;164: A1731–A44.

[53] Hanyu Y, Honma I. Rechargeable quasi-solid state lithium battery with organic crystalline cathode. Sci Rep. 2012;2:453.

[54] Huang W, Zhu Z, Wang L, Wang S, Li H, Tao Z, et al. Quasi-solid-state rechargeable lithium-ion batteries with a Calix[4]quinone cathode and gel polymer electrolyte. Angew Chem Int Ed. 2013;52:9162–6.

[55] Zhu Z, Hong M, Guo D, Shi J, Tao Z, Chen J. All-Solid-state lithium organic battery with composite polymer electrolyte and Pillar[5]quinone cathode. J Am Chem Soc. 2014;136: 16461–4.

[56] Lécuyer M, Gaubicher J, Barrès A-L, Dolhem F, Deschamps M, Guyomard D, et al. A rechargeable lithium/quinone battery using a commercial polymer electrolyte. Electrochem Commun. 2015;55:22–5.

[57] Bruce PG, Freunberger SA, Hardwick LJ, Tarascon J-M. Li–O2 and Li–S batteries with high energy storage. Nat Mater. 2012;11:19–29.

[58] Marmorstein D, Yu TH, Striebel KA, McLarnon FR, Hou J, Cairns EJ. Electrochemical performance of lithium/sulfur cells with three different polymer electrolytes. J Power Sources. 2000;89:219–26.

[59] Hassoun J, Croce F, Armand M, Scrosati B. Investigation of the O2 electrochemistry in a polymer electrolyte solid-state cell. Angew Chem Int Ed. 2011;50:2999–3002.

[60] Xia S, Wu X, Zhang Z, Cui Y, Liu W. Practical challenges and future perspectives of all-solid-state lithium-metal batteries. Chem. 2019;5:753–85.

[61] Balaish M, Peled E, Golodnitsky D, Ein-Eli Y. Liquid-free lithium–oxygen batteries. Angew Chem Int Ed. 2015;54:436–40.

[62] Zhu M, Schmidt OG. Tiny robots and sensors need tiny batteries – here's how to do it. Nature. 2021;589:195–7.

[63] Sun B, Liao IY, Tan S, Bowden T, Brandell D. Solid polymer electrolyte coating from a bifunctional monomer for three-dimensional microbattery applications. J Power Sources. 2013;238:435–41.

[64] Mindemark J, Sun B, Brandell D. Hydroxyl-functionalized poly(trimethylene carbonate) electrolytes for 3D-electrode configurations. Polym Chem. 2015;6:4766–74.

[65] Zadin V, Brandell D, Kasemägi H, Aabloo A, Thomas JO. Finite element modelling of ion transport in the electrolyte of a 3D-microbattery. Solid State Ionics. 2011;192:279–83.

[66] Grape U Recovery act – Solid state batteries for grid-scale energy storage (SEEO Final Technical Report). 2015.

[67] Ito S, Fujiki S, Yamada T, Aihara Y, Park Y, Kim TY, et al. A rocking chair type all-solid-state lithium ion battery adopting Li2O–ZrO2 coated LiNi0.8Co0.15Al0.05O2 and a sulfide based electrolyte. J Power Sources. 2014;248:943–50.

[68] Kobayashi Y, Seki S, Mita Y, Ohno Y, Miyashiro H, Charest P, et al. High reversible capacities of graphite and SiO/graphite with solvent-free solid polymer electrolyte for lithium-ion batteries. J Power Sources. 2008;185:542–8.

[69] Baudry P, Lascaud S, Majastre H, Bloch D. Lithium polymer battery development for electric vehicle application. J Power Sources. 1997;68:432–5.

[70] Lavoie P-A, Laliberté R, Dubé J, Gagnon Y Co-extrusion manufacturing process of thin film electrochemical cell for lithium polymer batteries and apparatus therefor. 2010.

[71] Tan DHS, Banerjee A, Chen Z, Meng YS. From nanoscale interface characterization to sustainable energy storage using all-solid-state batteries. Nat Nanotechnol. 2020;15:170–80.

[72] https://www.bluecar.fr/les-batteries-lmp-lithium-metal-polymere

5 Host materials

The literature on polymer electrolytes was for a long time dominated by a focus on poly-ethers, and in particular PEO. This largely started to change in the mid-2010s, when parallel work by several research groups began to open the field toward "alternative" host materials. It should be noted that some diversity existed in the field even before then, with important early work on, for example, polyesters, polyacrylonitrile (PAN) and poly (vinyl alcohol) (PVA), but this was not reflected in the overarching narrative of polymer electrolyte research. This chapter aims to reflect the considerable diversity of host polymers that have been used for SPEs up until the present time and accordingly highlight the unique characteristics of each class of materials. The division into subchapters is done primarily based on the coordinating moiety of each polymer class.

Polymers are, to varying degrees, characterized by long molecular chains, repeating structures and polydispersity that combine to give polymeric materials their unique properties that include high viscosities, slow diffusion and a tendency to resist crystallization. The length of chains in polymer materials may be described either by the degree of polymerization, which refers to the number of repeating units in a chain, or by the molecular weight of the chain. Because of the polydispersity of polymers (i.e., there is a distribution of chains with different lengths), any such value is by necessity an average over the entire system.

Apart from linear *homopolymers* comprising identical repeating units, other more complex architectures are also possible that include different degrees of branching, as well as *copolymers* formed from two or more comonomers being polymerized together to form chains comprising several different repeating units. Depending on the distribution of these repeating units, different copolymer architectures are possible, such as random, alternating, block and graft copolymers. The synthesis of block copolymers typically relies on sequential polymerization of individual comonomers using controlled polymerization processes that retain the end-group functionality, such as anionic polymerization, ring-opening polymerization (ROP) or atom transfer radical polymerization (ATRP). Branching increases the number of end groups in the system. Since the end groups tend to be structurally and functionally distinguished from the rest of the chain, and are characterized by faster dynamics as they can move more freely, highly branched systems gain properties that are dominated by the end groups and that may be distinctly different from linear chains of the same material.

Polymers with relatively short chain lengths may be referred to as *oligomers*. While it may be tempting to specify a fixed numerical cut-off molecular weight or chain length to define the division between oligomers and high-molecular-weight polymers, any such division will be inherently arbitrary. Much more useful is to consider the effects of molecular weight on the properties of the material to define this cut-off. As illustrated in Fig. 2.8, the onset of chain entanglements (leading to a well-defined increase in melt

https://doi.org/10.1515/9781501521140-005

viscosity) may be used as such a definition. Alternatively, the diminishing influence of end groups at higher molecular weight also serves as a divisive feature. This effect can, for example, be seen in the glass transition temperature (T_g), as described by the Flory–Fox equation:

$$T_g = T_{g,\infty} - \frac{K}{M_n} \tag{5.1}$$

For amorphous polymers, T_g is an important parameter as it describes the onset temperature of cooperative segmental motion in the system. As such, the T_g serves as a simple measure of the chain mobility of the material and becomes a parameter of high relevance for SPEs where the coupled ion transport mechanism dominates (see also Chapter 2 for a more in-depth discussion of these topics). In this context, Equation (5.1) describes how a higher concentration of end groups, which have a higher degree of freedom (more "free volume") and a higher mobility, lead to an overall lower T_g. This partially explains the higher ionic conductivity seen for oligomeric host materials, in combination with a transition toward vehicular ion transport. Of course, a sufficiently low molecular weight to give a substantially lower T_g is also detrimental to the mechanical properties of the material.

The effect of molecular weight on polymer chain dynamics is just one example of how structural and compositional factors may affect T_g. For random copolymers, the T_g becomes a weighted average of the glass transition temperatures of the respective homopolymers of the constituent monomers based on the weight fraction w_i of each component as described by the Fox equation:

$$\frac{1}{T_g} = \frac{w_1}{T_{g,1}} + \frac{w_2}{T_{g,2}} \tag{5.2}$$

Although polymers tend to resist full crystallization, it is not uncommon for polymers to exhibit semicrystallinity. While the presence of crystalline domains affects ionic movement directly, as discussed in Chapter 2, it also affects the T_g as the polymer ends become locked in crystalline structures, resulting in reduced chain dynamics for the relatively short free chain segments (essentially the same effect as in a cross-linked system).

5.1 Polyethers

5.1.1 Synthesis and structure of polyethers

Polyethers used as host materials for SPEs are invariably of the aliphatic type and, regardless of the length of the hydrocarbon chain between the ether oxygens, are synthesized through ROP. The archetypal PEO and similar oxyethylene-type polymers are obtained through ROP of the highly reactive three-membered cyclic ether ethylene oxide. At low molecular weights, and particularly with hydroxyl end groups, PEO is referred to as poly(ethylene glycol) (PEG), but is still synthesized from the same ethylene oxide monomer. As the ethylene oxide monomer is a toxic gas under standard conditions, the synthesis of PEO and analogues is normally not performed in a research lab setting. Commercial PEO may contain up to 1,000 ppm of *tert*-butylated hydroxytoluene (BHT) added as an antioxidant. This can be removed by, for example, Soxhlet extraction with hexanes in order to prevent its impact on the electrochemical performance of the material [1], although the standard practice in the literature is rather to leave any additives in the material.

Ethylene oxide and other three-membered cyclic ethers (known as *epoxides*) can be polymerized using both cationic and anionic mechanisms. The anionic route is preferable, since it enables better control and higher molecular weights. Compared to ethylene oxide, propylene oxide is much more difficult to polymerize with a high level of control even with the anionic approach due to extensive chain transfer reactions [2]. Consequently, only low molecular weights of poly(propylene oxide) are readily accessible. More complex functionalities may also be accessed by the use of glycidyl ether monomers, as illustrated in Fig. 5.1. Based on the same epoxide functional group, they can also be polymerized through the same pathways.

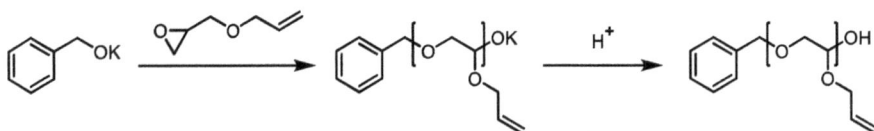

Fig. 5.1: Example of ROP of allyl glycidyl ether initiated by potassium benzoxide and terminated by a proton source according to Barteau et al. [3].

Larger cyclic ethers, such as oxetane and tetrahydrofuran, may be polymerized cationically (Fig. 5.2). As the rings grow larger, they become progressively more difficult to polymerize because of diminishing ring strain. This is reflected by the heat of polymerization (Tab. 5.1), which roughly follows the (negative) strain energies of the respective monomers. Whereas poly(trimethylene oxide)/polyoxetane and poly(tetramethylene oxide)/polytetrahydrofuran are both feasible, at a ring size of six (tetrahydropyran, 1,4-dioxane) there is no longer any sufficient driving force for ring-opening and polymerization is not viable.

For architectures where oxyethylene-based chains are grafted as side chains to other polymeric backbones to form comb-type polymers, the most straightforward synthesis routes involve PEG chains or glymes and *grafting to* or *grafting through* approaches. The "grafting to" approach is useful for functionalizing, for example, polyphosphazene backbones (Fig. 5.3) whereas the "grafting through" approach can be illustrated by the polymerization of a functionalized dichlorosilane monomer to create a polysiloxane with glyme side chains as shown in Fig. 5.4. Flexible grafted side chains can also be incorporated in block copolymer architectures by controlled radical polymerization methods in a "grafting through" approach, as illustrated in Fig. 5.5.

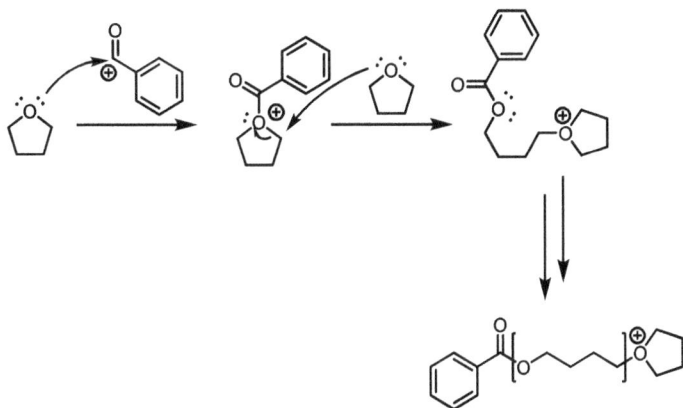

Fig. 5.2: Example of ROP of tetrahydrofuran initiated by the benzoyl cation as described by Alamgir et al. [4]. The antimony hexafluoride counterion has been omitted for clarity. Termination by nucleophilic agents removes the activated tetrahydrofuran end-group.

Tab. 5.1: Comparison of the heat of polymerization of cyclic ethers depending on ring size. Data from [5].

Monomer	Ring size	$-\Delta H_p$ (kJ mol^{-1})
Ethylene oxide (oxirane)	3	127.3
Trimethylene oxide (oxetane)	4	81.8
1,3-Dioxolane	5	50
Tetrahydrofuran (oxolane)	5	22
Tetrahydropyran (oxane)	6	1.5
Hexamethylene oxide (oxepane)	7	38.6
1,3-Dioxepane	7	53.9

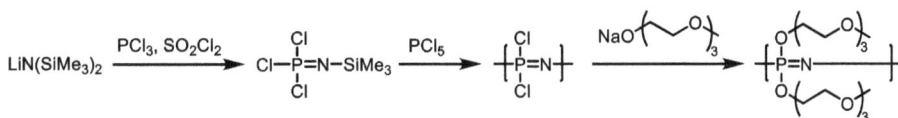

Fig. 5.3: Synthesis and functionalization of a polyphosphazene backbone with glyme side chains as described in [6].

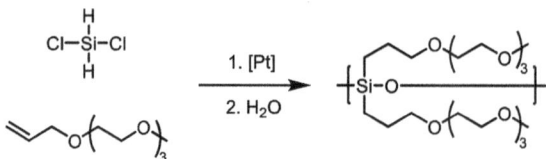

Fig. 5.4: Synthesis and polymerization of a glyme-functionalized dichlorosilane as described in [7].

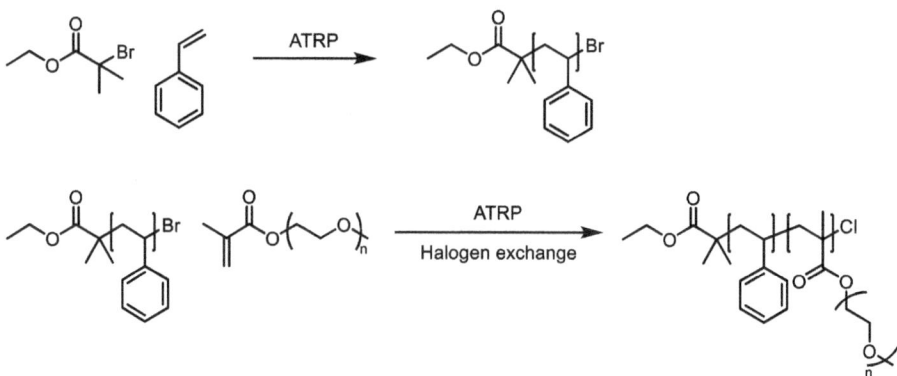

Fig. 5.5: Synthesis of a polystyrene–poly(oxyethylene methacrylate) block copolymer by ATRP as described in [8]. Here, $n = 9$.

5.1.2 PEO and oxyethylene-based polyethers

The historically most important host material for SPEs is undoubtedly PEO, and it continues to be relevant in current research. It is also from studies on PEO that most of what is currently known about ion conduction in SPEs is known. SPEs based on PEO and other host materials characterized by ion coordination to oxyethylene repeating units have recently been covered in several reviews [9].

One important reason for the dominance of PEO as a host material in the scientific literature is related to its perhaps unique ability to solvate Li$^+$ ions and thereby dissolving lithium salts. Conventional wisdom relates the solvation ability of a solvent to its Lewis basicity, commonly described by its donor number [10]. While this parameter is generally higher for ethers than for, for example, carbonates or esters (indicating stronger ion solvation by ethers), this trend is not necessarily followed when considering specific cations. When considering the particular case of Li$^+$ solvation, stronger coordination by carbonyl oxygens in carbonates, esters and ketones is instead observed [11]. The situation is, however, reversed in the presence of the *chelating* effects in multidentate ether solvents such as PEO. Particularly the two-carbon spacing between successive oxygens in oxyethylene-based polyethers turns out to be optimal for Li$^+$ and makes polymers based on the $-CH_2CH_2O-$ structural unit much better solvents for Li salts than polymers based on either one-carbon ($-CH_2O-$) or three-carbon ($-CH_2CH_2CH_2O-$) repeating units [12]. It has been found that, in an elec-trolyte consisting of up to 90 mol% propylene carbonate with the rest being tetraethy-lene glycol dimethyl ether (TEGDME), most Li^+ remains coordinated by ether oxygens from the TEGDME [13]. Moreover, the exchange of solvent molecules in oligo(ethylene oxide):Li^+ is notably slow compared to small-molecule carbonate electrolyte solvents [14], and complexes of suitably sized glymes with Li salt show remarkable stability, behaving as single coherent entities referred to as "solvate ionic liquids" [15].

The oxyethylene repeating units render PEO basically a high-molecular-weight glyme, and PEO exhibits similar chelating effects in ion solvation. Furthermore, PEO has a comparatively low T_g (−60 °C), which is another prerequisite for comparatively fast ion conduction. Importantly, PEO is crystalline to a large degree; pure PEO crystal-lizes to 75–80% at room temperature [16]. Although the degree of crystallinity dimin-ishes with the addition of salt, the crystallinity of PEO severely restricts ion transport below the melting point (60 °C). A possible exception is certain crystalline phases of PEO with Li and Na salts where significant crystallinity has been reported [17–19], al-though it should be noted that there is still some controversy surrounding the idea of fast ion conduction in crystalline PEO:salt phases [20, 21].

The low T_g of PEO translates into fast ion conduction in the amorphous state above the melting point, as seen in Fig. 5.6. As illustrated in Fig. 5.7 for a series of PEGs, the conductivity is dependent on the molecular weight up until the high-molec-ular-weight limit [22]. This can be related both to the decrease in T_g, described by the Flory–Fox equation, and a transition in ion transport mechanism from coupling to segmental motions toward vehicular transport with decreasing molecular weight. This also affects T_+, which decreases with increasing molecular weight to a stable high-mo-lecular-weight plateau between 0.1 and 0.2 in the case of Li^+ conduction [22].

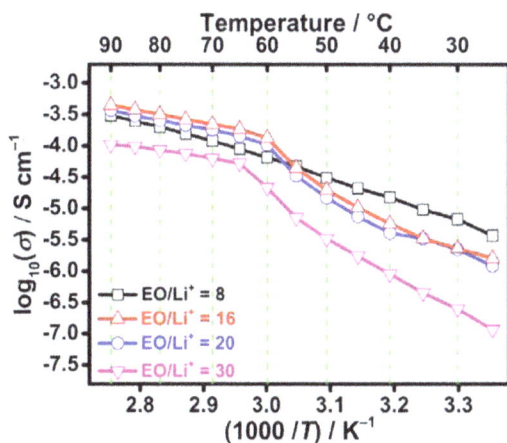

Fig. 5.6: Arrhenius plot of conductivity for PEO:LiTFSI electrolytes. Reprinted with permission from [23]. Copyright 2016 American Chemical Society.

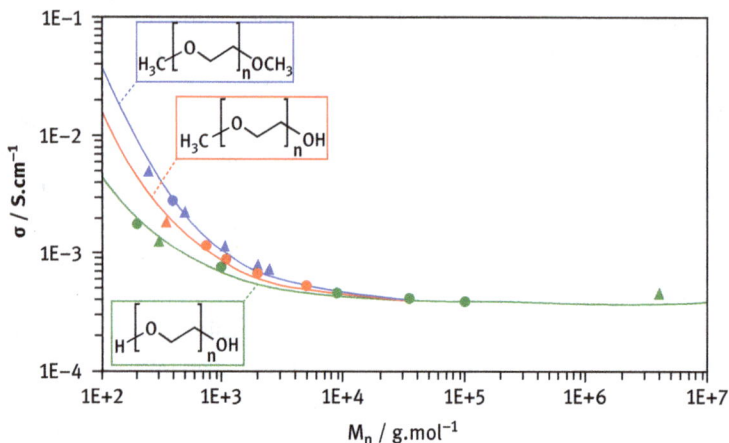

Fig. 5.7: Variation of ionic conductivity with molecular weight for different varieties of PEG$_{25}$LiTFSI electrolytes at 60 °C. Adapted from [22], Copyright 2012, with permission from Elsevier.

These low transference numbers are typical for Li$^+$ conduction in PEO-based electrolytes and the poor cation mobility is even more pronounced with Mg^{2+} [24]. This effect can be traced back to the excellent complexation of cations by oxyethylene chains; while this facilitates salt dissolution, it also restricts the release of the cation as new coordination environments are presented by the polymer segmental motion. The result is very stable coordination structures that counteract the effects of the low T_g and fast polymer chain dynamics, creating the unfortunate situation that the anion is much more mobile than the cation [25]. This leads to the awkward situation that the very property that gives PEO its remarkable ability to solvate Li$^+$ and other cations also serves to severely restrict their movements in the polymer matrix.

The electrochemical stability of PEO in realistic battery systems is not very much explored, but the SEI layer formation on the anode side is primarily considered to consist of products related to salt decomposition and water traces in the hygroscopic PEO [26]. There exist so far only very limited studies of the degradation layer characteristics on the Li-battery cathode side for PEO-based electrodes [27], but short-chain ether homologues show inferior stability at higher potentials compared to liquid carbonates [28], which indicates that polyethers might suffer extensive decomposition on this electrode side, especially for high-voltage electrodes. On the other hand, polyethers display apparent functionality with the Li-metal electrode. Therefore, as described in Chapter 4, a typical functional battery construction is the Li | PEO:salt | LFP cell.

The phase diagrams of PEO:salt systems are often complex, with multiple phases present in temperature ranges of relevance for battery applications (Fig. 5.8). The conductivity behavior can therefore be unpredictable. Above the melting point, where the ionic conductivity is sufficient for battery operation, the mechanical stability is

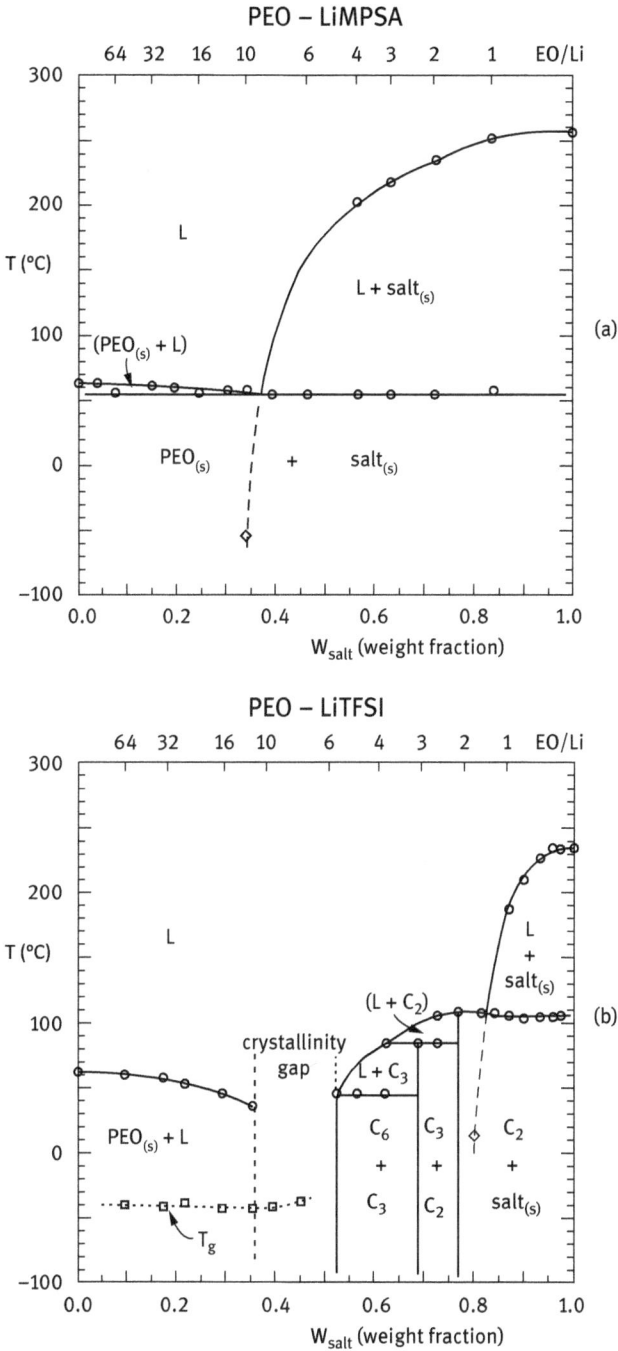

Fig. 5.8: Phase diagrams of (a) PEO:LiMPSA and (b) PEO:LiTFSI. Adapted with permission from [35]. Copyright 1994 American Chemical Society.

unfortunately insufficient in the pure electrolyte material. This is basically the classic problem of the coupling of ionic conductivity to chain flexibility, and can be tackled by introducing the ion-coordinating motif in more complex polymer architectures such as block copolymers [29–31], graft copolymers [32, 33] or cross-linked systems [34]. These may be motivated by a separation of load-bearing and ion-conducting functions (block copolymers, cross-linked systems) or improvement of the chain flexibility for increased conductivity. Graft copolymers can fall into either category, depending on the properties of the main polymer backbone.

With block copolymers, a whole new morphological dimension opens up as the phase separation between blocks can lead to a range of complex microstructures that affect the ion transport in the system [36–38]. Block copolymer architectures, similar to cross-linking, will also lead to a local stiffening of the material because of the anchoring of the chains. If the ion-conductive groups are instead introduced as side chains in graft copolymers, the end groups are left free to move, resulting in high segmental mobility of the side chains while the main chain may infer mechanical stability [39]. The ion-coordinating ether groups can also be grafted onto highly flexible backbones, such as polysiloxanes [40–42] or polyphosphazenes [43] to boost the overall molecular flexibility. These systems are both based on backbones with low barriers for bond rotations in the main chain, resulting in high chain flexibility and low glass transition temperatures. With short oligo(ethylene oxide) side chains, the overall high mobility of the system can be retained and crystallization prevented; for polysiloxanes, an optimal chain length equivalent to 6 oxygens has been reported [44]. High conductivities can indeed be attained in such ultra-flexible systems – 4.5×10^{-4} S cm^{-1} at ambient temperature has been reported for polysiloxanes [44] – although the slippery polymer chains provide little mechanical stability. This is further exacerbated at the often low molecular weights seen in these systems, leaving the electrolytes in a viscous liquid state unless stabilized by, for example, cross-linking.

As analogues to PEO, the perfluorinated equivalents to this host polymer have also been investigated [45–48]. While PEO electrolytes are considerably less flammable than liquid electrolytes based on low-molecular-weight solvents, these perfluoropolyethers are truly nonflammable [47]. These oligomeric perfluoroethers have extremely low glass transition temperatures, but unfortunately have a poor ability to dissolve Li salt. This results in the ionic conductivity being limited by the maximum salt concentration and reaches a magnitude of 10^{-5} S cm^{-1} at 30 °C for the shortest oligomer [47]. It should be noted that these electrolytes are all liquid because of the limited molecular weights and high molecular flexibility. Mechanical stabilization by cross-linking results in diminished ionic conductivity by two orders of magnitude [45]. Interestingly, the perfluoropolyether electrolytes show much higher T_+ than comparable PEO electrolytes [47, 48]. They also appear to act stabilizing to the SPE/Li metal interface and are capable of suppression of Al current collector corrosion in battery cells [49].

5.1.3 Other polyethers

The crystallinity of PEO is widely recognized as a key issue that limits electrolyte performance. This can be countered by using other polyethers, that show similarly good ion solvation, but that do not crystallize. Examples of such polymers are PPO [50–54] and poly(allyl glycidyl ether) (PAGE) [3]. Polymerization of 1,3-dioxolane also leads to an alternating oxymethylene/oxyethylene copolymer that, unlike PEO, becomes amorphous on addition of salt and therefore shows improved ionic conductivity at room temperature [4].

PPO does not solvate Li^+ ions as efficiently as PEO and there is a low degree of salt dissociation in PPO matrices [55]. This results in lower ionic conductivity being observed for PPO electrolytes despite having similar ionic mobility to PEO electrolytes, clearly indicating differences in free ion concentration [56].

Interestingly, despite being both fully amorphous and having a lower T_g than PEO, electrolytes based on PAGE only show higher ionic conductivity at temperatures below the melting point of PEO. When compared in the temperature range where PEO is also amorphous, PEO shows the highest conductivity, indicating an inherently better ion transport ability of the polyether without the plasticizing allyl ether side chains [3]. In PAGE and similar systems the side chains will limit how the polymer chains can come into close proximity to each other, thereby preventing the formation of suitable coordination sites and reducing the solvation site connectivity [57]. In fact, among host materials with oxyethylene moieties as the coordinating groups, PEO appears to have the optimal structure for facile ion movement [58].

Apart from the crystallinity, strong ion–polymer interactions constitute the other major caveat with oxyethylene-based host materials. This can also be addressed by structural modifications, such as replacing every other oxyethylene repeating unit with a trimethylene oxide repeating unit to weaken the ion binding, leading to faster Li^+ transport according to MD simulations [59]. Similar effects can also be attained with other polyether hosts, such as polytetrahydrofuran. With a low T_g, polytetrahydrofuran shows slightly improved room temperature conductivity compared to PEO (Fig. 5.9), but the most notable difference is the weaker cation coordination, leading to an improvement in T_+ [4, 60]. Originally dismissed because of low thermal stability [61, 62], cross-linked membranes of polytetrahydrofuran were recently developed that show sufficient stability for practical applications [60].

Related to this class of polyethers are also the cyano-functional polyoxetanes and polymethacrylamides PCEO, PCOA and PMCA described by Tsutsumi et al. that essentially constitute hybrids of polyethers and polynitriles [63–65]; these will be discussed further as polynitriles in Section 5.3.

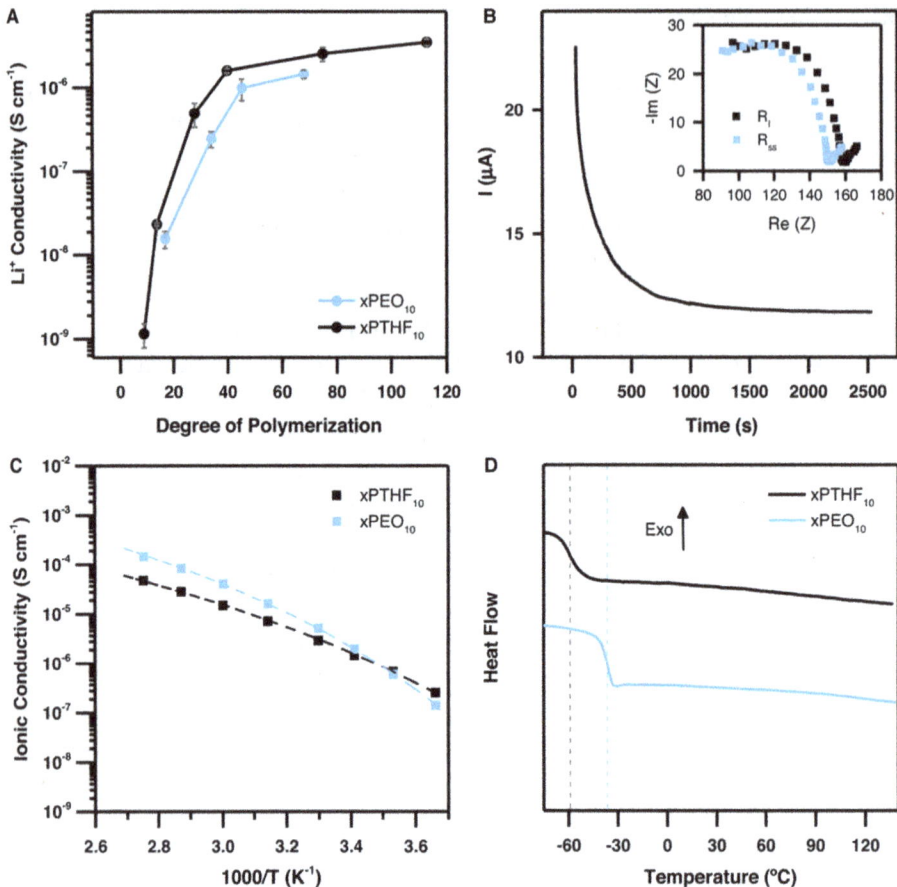

Fig. 5.9: Properties of cross-linked polytetrahydrofuran:LiTFSI electrolytes compared to a cross-linked PEO reference. Reproduced with permission from [60]. © 2018 WILEY-VCH.

5.1.4 Application of polyethers in batteries

Given the ubiquity and long-term reliance of PEO as a host material for SPEs, it is no surprise that much of the efforts into building SPE-based all-solid-state Li batteries have been based on PEO-based electrolytes. To date, PEO is also more or less the only host material for SPEs that has seen some degree of commercial success. The crystallinity of PEO precludes its use at temperatures below its melting point. Consequently, high-temperature operation at 80–100 °C is typically employed to improve the ion transport to the point where sufficient currents can be achieved. Battery operation with PEO-based SPEs at room temperature instead requires structural modifications [66]. One strategy to enable battery cycling at room temperature is to disrupt crystallinity by copolymerization [66, 67]. The poor performance of PEO at room temperature

is not purely related to crystallinity; electrolytes based on random PEO/PPO copolymers also do not display satisfactory performance at ambient temperature, suggesting
more fundamental ion transport limitations [68]. Faster ion transport for successful
room temperature battery operation has been achieved using high-molecular-weight
copolymers with highly flexible PEO- and polysiloxane-based side chains [69] or flexible polysiloxane backbones with grafted short PEO chains [70]. It should be noted that
these strategies often compromise the mechanical properties of the material, effectively preventing the electrolytes from functioning as reliable solid separators without
mechanical stabilization by a separate polymer membrane or similar [71].

When used in their proper habitat – that is, at elevated temperatures and modest
cycling rates – polyether-based SPEs perform well, particularly with the uncomplicated
cathode material LFP, with stable long-term cycling. This is illustrated in Fig. 5.10,
which shows the cycling performance of a Li half-cell based on a copolymer of ethylene
oxide, 2,2-methoxyethoxyethyl glycidyl ether and allyl glycidyl ether (82/18/1.7 ratio)
[72]. In this material, the glycidyl ethers serve to disrupt crystallinity, improve the chain
(and ion) dynamics and enable cross-linking for mechanical stability. While this (and
most other examples of PEO-based batteries) constitutes a half-cell with metallic Li as
the anode, similar SPEs have also been successfully applied in full-cell configurations.
In the example shown in Fig. 5.11, a high-molecular-weight poly(ethylene oxide-co-2,2-
methoxyethoxyethyl glycidyl ether) (88/12 ratio) with LiTFSI salt was impregnated into
the electrode and combined with a top layer of cross-linked poly(ethylene oxide-co-2,2-
methoxyethoxyethyl glycidyl ether-co-allyl glycidyl ether) for mechanical stabilization
and thereafter cycled in a SiO/graphite ‖ LFP cell configuration [73].

When it comes to more demanding cathode materials that operate at higher potentials, such as $LiCoO_2$ and NMC, these are generally more problematic for stable
battery operation using polyethers. The traditional PEO:LiTFSI cannot withstand
the high operating potential of these electrode materials, particularly at the elevated
temperatures necessary for sufficient ionic conductivity. This is manifested as voltage instabilities and poor cyclability in, for example, Li ‖ NMC622 half-cells. This
can be improved with more complex nanocomposite materials to the point where
reversible cycling can be achieved, but with steadily declining capacities and poor
long-term stability [74]. Recent data also highlights problems with dendritic lithiation that can be overcome through the use of full cells or thicker electrolyte membranes to improve cyclability versus NMC622 [75], but ultimately without long-term
capacity retention. Another viable strategy to prevent electrochemical degradation
at the electrode–electrolyte interface is to apply a protective coating to the cathode
material surface. Fig. 5.12 shows the effect of introducing a sacrificial buffer layer
consisting of sodium carboxymethyl cellulose (CMC) on the surface of NMC111 particles for the cycling of prototype half-cells [76].

Before the commercialization of the liquid-electrolyte Li-ion battery based on a
carbonaceous anode, PEO- and other polyether-based SPEs were extensively re-
searched and developed for the practical realization of Li batteries based on metallic

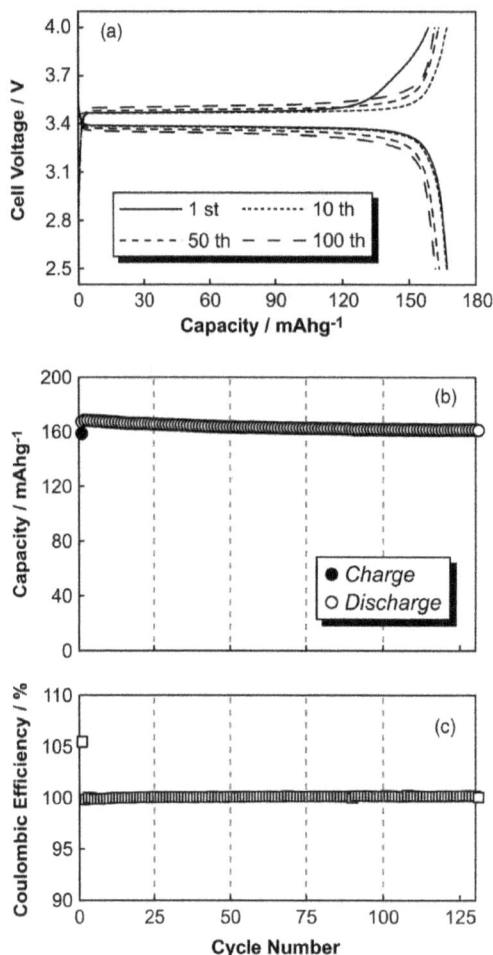

Fig. 5.10: Cycling performance of a poly(ethylene oxide-*co*-2,2-methoxyethoxyethyl glycidyl ether-*co*-allyl glycidyl ether):LiTFSI electrolyte in a Li || LFP cell configuration at C/8 and 333 K. Reprinted from [72]. © IOP Publishing. Reproduced with permission. All rights reserved.

Li anodes. Consequently, all-solid-state Li batteries based on polyether SPEs are well-developed and provide excellent functionality in prototype cells. In contrast, Na-based systems are much more challenging, particularly if metallic Na is used as the anode. While considerably fewer than for their Li counterparts, there are nevertheless examples of sodium battery prototypes implementing polyether-based SPEs, both older [77, 78] and more recent [79, 80]. Metallic sodium has, for example, been combined with PEO:NaClO$_4$, reinforced with Na-carboxymethyl cellulose, and TiO$_2$ and NaFePO$_4$ cathodes at 60 °C [79]. Another example is shown in Fig. 5.13, where a PEO:NaFSI electrolyte was used in combination with sodium metal and Na$_{0.67}$Ni$_{0.33}$Mn$_{0.67}$O$_2$ or Na$_3$V$_2$(PO$_4$)$_3$@C [80]. These examples highlight the research

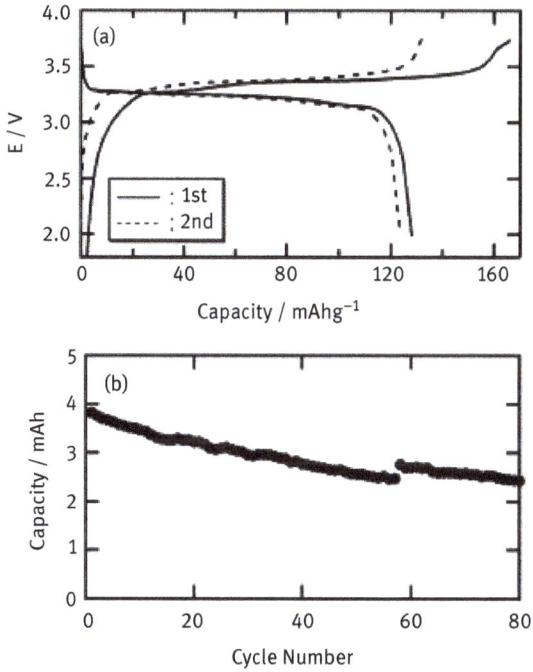

Fig. 5.11: Cycling performance of a SiO/graphite || LFP full cell based on a combination of polyether electrolytes. Adapted from [73], Copyright 2008, with permission from Elsevier.

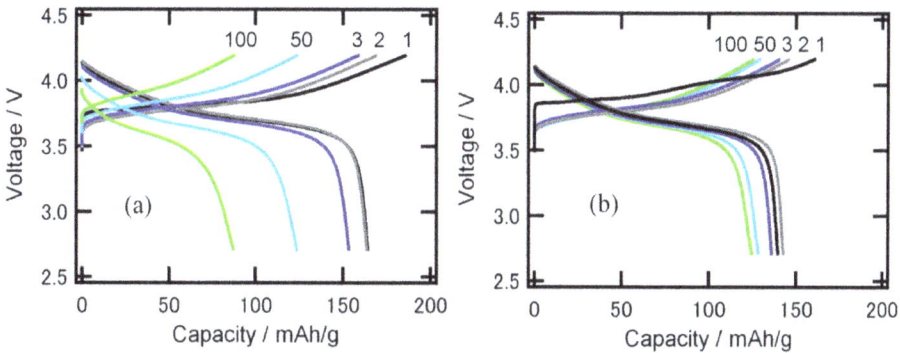

Fig. 5.12: Voltage profiles of selected cycles for Li || NMC111 battery cells with a polyether electrolyte, comparing (a) bare NMC111 particles with (b) CMC-covered NMC111. Reprinted with permission from [76]. Copyright 2013 American Chemical Society.

efforts into the application of polyether-based SPEs for Na batteries, although it should be noted that the obstacles for full practical realization of these technologies have not yet been overcome.

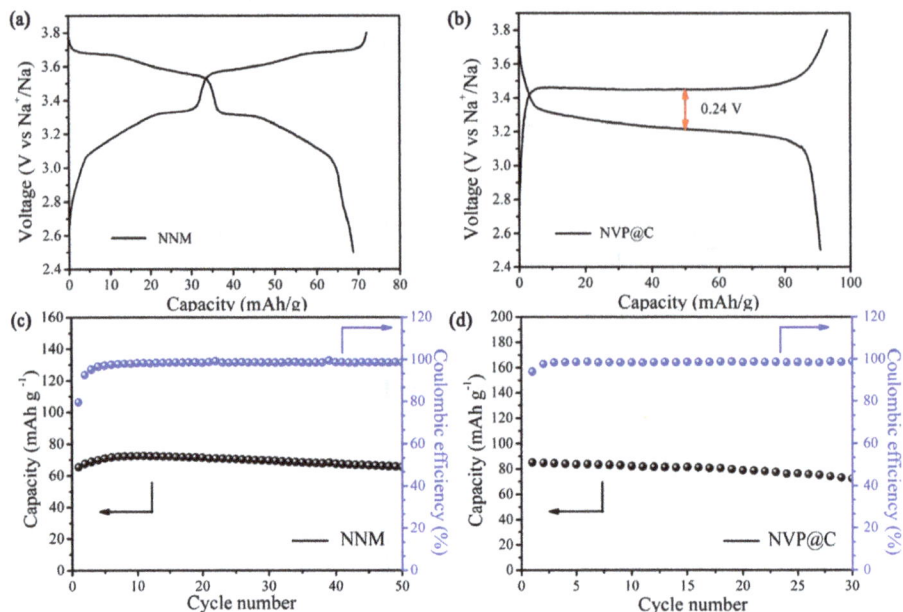

Fig. 5.13: Cycling performance of (a, c) Na || Na$_{0.67}$Ni$_{0.33}$Mn$_{0.67}$O$_2$ (NNM) and (b, d) Na || Na$_3$V$_2$(PO$_4$)$_3$@C (NVP@C) cells with a PEO:NaFSI electrolyte at 0.2 C and 80 °C. Reproduced with permission from [80]. © 2016 Wiley-VCH.

5.2 Carbonyl-coordinating polymers

In contrast to the dominance of polyethers in SPEs, conventional liquid electrolytes for Li-ion batteries are more typically based on molecules that solvate cations by means of carbonyl groups, more specifically organic carbonates. These are technically esters of carbonic acid, but are often distinguished as a separate (sub)class. The polymeric equivalents – polycarbonates and polyesters – have also been extensively explored for use as host materials in SPEs. Very recently, polyketones have also been introduced as a third type of carbonyl-coordinating host materials, and will be discussed along with polycarbonates and polyesters within this section.

The carbonyl group of polycarbonates, polyesters and polyketones is useful as a sensitive probe of cation coordination, as the C=O stretch vibration is a prominent feature in the IR spectra of these polymers. On coordination, the frequency of this vibration shifts toward lower wavenumbers, as illustrated in Fig. 5.14. Comparison of the integrals of the resulting peaks gives the fraction of carbonyl groups involved in ion coordination, which can be used to calculate the coordination number (with respect to carbonyl groups):

$$n_{C=O} = \chi \times n \tag{5.3}$$

where χ is the fraction of ion-coordinating carbonyl groups and n is the carbonyl:Li$^+$ molar ratio [81]. This can be used to closely follow the coordination in carbonyl-based systems and, when combined with measurements of ion pairing, may give important information about the evolution of the coordination structures with salt concentration. This is particularly relevant in systems with the possibility of mixed coordination between several different coordinating functional groups.

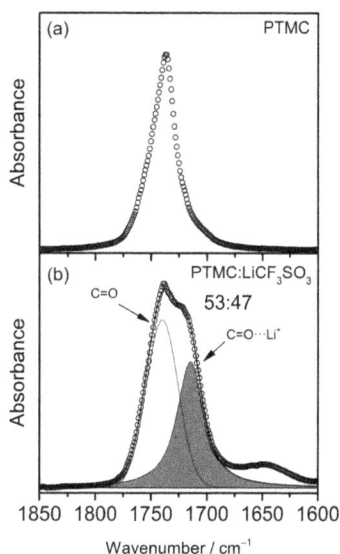

Fig. 5.14: Detection of carbonyl coordination using FTIR (a). The annotated numbers in (b) refer to the ratio of non-coordinating to coordinating carbonyls. Data from [81].

For high ionic conductivity, only those polymers that have aliphatic backbones are of relevance, as the slow chain dynamics and high T_g associated with aromatic backbones limit ion transport. This excludes the ubiquitous commodity polymers of this class such as poly(ethylene terephthalate) and poly(bisphenol A carbonate).

5.2.1 Synthesis of carbonyl-coordinating polymers

Polycondensation of difunctional alcohols with difunctional carboxylic acids or acid derivatives works as a general method for the synthesis of polyesters and polycarbonates. In the case of polycarbonates, carbonic acid cannot be used for the synthesis and instead esterification with phosgene or derivatives thereof may be used. A safer alternative is to utilize transesterification of low-molecular-weight organic carbonates, for example dimethyl carbonate, as illustrated in Fig. 5.15a [82]. By driving off the liberated alcohol at high temperature and vacuum, the product can be pushed toward higher molecular weights. While polycondensation is a straightforward synthesis method that is versatile in terms of the accessible structures, it is often limited in

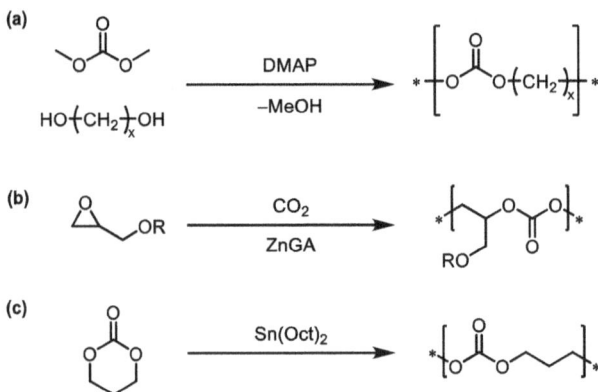

Fig. 5.15: (a) Synthesis of polycarbonates by polycondensation catalyzed by 4-dimethylaminopyridine (DMAP). (b) Synthesis of functional PEC-type polycarbonates by copolymerization of glycidyl ethers with carbon dioxide catalyzed by zinc glutarate (ZnGA). (c) Synthesis of poly(trimethylene carbonate) by ROP of trimethylene carbonate catalyzed by tin(II) 2-ethylhexanoate (Sn(Oct)$_2$).

terms of what molecular weights that can be attained and offers no direct control over either end-groups or the molecular weight. Moreover, polycondensation cannot be used to synthesize polycarbonates with two carbon atoms in the main chain of the repeating unit, as the difunctional alcohol in these cases instead tends to ring-close to the five-membered cyclic carbonate.

For better control over molecular weights, architectures and end-groups, as well as access to materials that are inaccessible with polycondensation, ROP is a useful method that has found much application for polycarbonate and polyester synthesis. Analogous to the cyclic ethers, the ring strain as a driving force for polymerization varies with the ring size, but has a minimum at five- rather than six-membered rings for simple cyclic esters (Tab. 5.2). Cyclic carbonates follow a similar pattern, with six- and seven-membered monomers being readily polymerizable, whereas the five-membered rings, for example ethylene carbonate, can only be made to undergo polymerization under much harsher conditions [83]. Under these conditions, the polymerization is accompanied by decarboxylation to form random polycarbonate/polyether copolymers [84]. To synthesize high-molecular-weight poly(ethylene carbonate) (PEC), poly(propylene carbonate) (PPC) and similar polycarbonates with two carbons in the main chain, alternating copolymerization of epoxides with carbon dioxide catalyzed by, for example, zinc glutarate, is used instead (Fig. 5.15b) [85].

The six-membered cyclic carbonate platform, on the other hand, is an excellent basis for ROP synthesis of polycarbonates based on the poly(trimethylene carbonate) (PTMC) backbone. Polymerization can be done under both cationic and anionic conditions [83], as well as using organocatalysts [86], but the most straightforward route to high-molecular-weight materials is probably tin(II)-catalyzed ROP following a coordination–insertion mechanism (Fig. 5.15c). If desired, the molecular weight can be readily controlled by means of introducing a protic initiator, such as an alcohol. The same approach can also be used for cyclic esters (lactones), enabling also facile synthesis of polyester–polycarbonate copolymers [87].

Tab. 5.2: Comparison of the heat of polymerization of cyclic esters depending on ring size. Data from [5].

Monomer	Ring size	$-\Delta H_p$ (kJ mol^{-1})
β-Propiolactone	4	75
γ-Butyrolactone	5	−5
δ-Valerolactone (pentanolactone)	6	10.5
ε-Caprolactone	7	17

The third class of carbonyl-coordinating host materials, *polyketones*, has only recently been introduced for SPEs, and requires more specialized synthesis methods. One example is the iterative approach by Manabe et al., where activated monomers are added consecutively to an activated growing chain [88], thereby allowing the build-up of successively larger oligomers, as illustrated in Fig. 5.16.

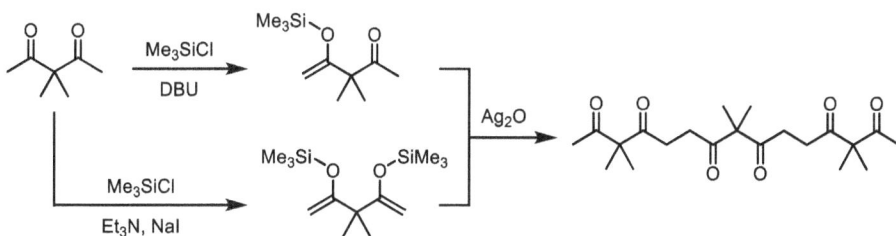

Fig. 5.16: Synthesis of polyketone oligomers using an iterative approach.

5.2.2 Polycarbonates

Considering the ubiquity of liquid organic carbonate solvents in Li-ion batteries, it is natural that a lot of research attention is currently being directed also toward polycarbonates as host materials for SPEs. Carbonates can coordinate Li$^+$, Na$^+$ and similar metal cations through their carbonyl oxygens. In liquid organic carbonates, the alkoxy oxygens next to the carbonyl group (often referred to as "ether oxygens") may also be involved in coordination to some extent, but this is rarely observed in polycarbonates [57, 89], possibly for steric reasons.

A key aspect to understanding the properties of polycarbonates as host materials is their relatively weak ion–polymer interactions. Owing to the weak coordination strength of the Li$^+$ cation by the carbonyl group in polycarbonates [25], for example, these are generally characterized by high T_+; values as high as 0.83 have been reported for PEC$_1$LiTFSI [90] and 0.80 for PTMC$_8$LiTFSI [89]. This is in striking contrast to traditional PEO-based systems and is also superior to typical results from polyester systems. The contrast between oxyethylene and carbonate coordination is also highlighted by the preference of Li$^+$ for coordination by sequential oxyethylene moieties. This can clearly be seen if glyme-like moieties or oligoethers are either tethered as side chains to a polycarbonate backbone or introduced as a linear backbone block. While this may improve the molecular dynamics, such ether features will, if these segments are sufficiently long to form stable coordination structures, take complete precedence over the ion transport properties as they will dominate the ion coordination in the system [91–93]. Consequently, T_+ will sharply drop in such systems. This is not the case if only individual ether oxygens are present, preventing chelating structures to be formed. In such cases, while the ether oxygens do take part in Li$^+$ coordination, the cations do not show preference for ether coordination.

As already mentioned, aliphatic backbones are preferred for sufficient molecular flexibility to allow for fast ion dynamics. The simplest such polymer is poly(ethylene carbonate) (PEC), with two methylene groups in the main chain of each repeating unit. Whereas PEC is synthesized from alternating copolymerization of ethylene oxide with carbon dioxide, the use of other epoxide monomers will result in a diversity of side chain-functional PEC derivatives, as shown in Fig. 5.17. PEC is amorphous with a relatively high T_g of 9 °C and behaves similarly to polyether SPEs when combined with LiCF$_3$SO$_3$ or LiClO$_4$ to form SPEs, showing the typical stiffening of the material as a consequence of the transient physical cross-links introduced by

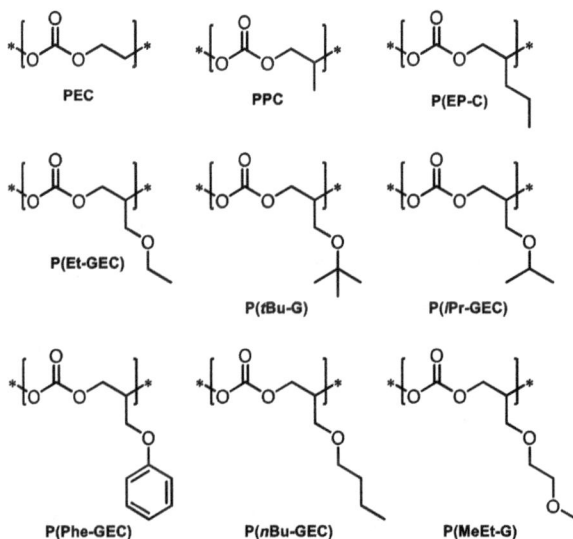

Fig. 5.17: A selection of PEC-type polycarbonates used as host materials for SPEs.

the cation coordination in the system, thereby hindering ion transport at high salt concentrations. With LiBF$_4$, LiBETI, LiTFSI or LiFSI, however, the salt instead has a plasticizing effect, resulting in SPE systems that, while displaying relatively poor conductivity at low salt concentrations, show comparatively fast ion conduction in the high-salt regime [94, 95]. In the PEC:LiFSI system, for example, the ionic conductivity reaches up to 4.0×10^{-4} S cm^{-1} at 40 °C for PEC$_{0.53}$LiFSI. As illustrated in Fig. 2.11, there is thus a very clear contrast between this system and equivalent PEO:LiFSI electrolytes, where the latter shows a distinct conductivity maximum at much lower salt concentration. This increase in molecular and ion dynamics is also accompanied by a predictable deterioration of the mechanical properties.

The behavior of PEC electrolytes at high salt concentrations has been attributed to the plasticizing effects of salt aggregates formed when entering the PISE regime, which facilitate enhanced rotational mobility in the polymer chains as well as reduce intramolecular interactions in the PEC chains. In the PEC:LiFSI system such ion aggregation can be observed already at moderate salt concentrations, but ionic aggregates become the dominant salt species at concentrations exceeding 50 mol% relative to the carbonate groups. This strongly suggests a percolation-type ion transport mechanism as the reason behind the fast ion conduction, but the conduction mechanism shows elements of both PISE-type conduction and conduction coupled to segmental motions. Importantly, the conductivity rises as the T_g decreases, indicating no major contributions from decoupled ion transport. Although this has been more thoroughly studied for the PEC:LiFSI system, the suggested mechanism very likely applies to PEC:LiTFSI electrolytes as well, given the similarities between LiFSI and LiTFSI.

The ionic conductivity of PEC can be increased by randomly incorporating oxyethylene units in the main chain to obtain the polycarbonate/polyether hybrid P(EC/EO) [96]. With this arrangement of ether oxygens, the ethers are prevented from forming chelating structures with Li$^+$, and the result is a cation transference number of 0.66 for P(EC/EO):LiTFSI. While this is lower than for PEC electrolytes systems, it is much higher than what is seen for PEO:LiTFSI.

The weak interactions between cations and polymer chains in PEC electrolytes have also been demonstrated to lead to improved electrochemical stability when the salt concentration is increased. In both PEC:LiTFSI and PEC:LiClO$_4$ at high concentrations, improved oxidation resistance and inhibition of aluminum corrosion have been noted [97].

High ionic conductivity has also been reported for PPC. Similar to PEC, this polymer has a high T_g of 24 °C that decreases to 5 °C when combined with 23 wt% LiTFSI. When supported by a cellulose membrane for mechanical stability, this electrolyte has a reported conductivity of 3.0×10^{-4} S cm^{-1} at 20 °C [98]. Data for other salt concentrations confirm the trend of decreasing T_g with increasing salt concentration in the same fashion as for PEC [99]. With 18 wt% KFSI salt, 1.36×10^{-5} S cm^{-1} at 20 °C has been reported [100]. When in contact with Li metal, PPC degrades to micromolecular segments that can infiltrate the interface and swell the electrolyte, reducing both bulk and interfacial resistances. In addition, the polymer undergoes depolymerization to propylene carbonate [101]. This is a testament to the exceptional stability of five-membered cyclic carbonates. Consequently, the same should also apply to PEC, which has a known thermal instability. In fact, this has recently been suggested to be part of the origin of the peculiar properties of electrolytes based on PEC or PPC. As summarized in Fig. 5.18, there is a tendency for significant amounts of solvents to remain in the system at high salt concentrations, while the thermal instability instead leads to significant depolymerization under conditions sufficient to obtain fully dried SPEs [102]. There is thus an apparent risk of inadvertent plasticization by low-molecular-weight compounds in these systems.

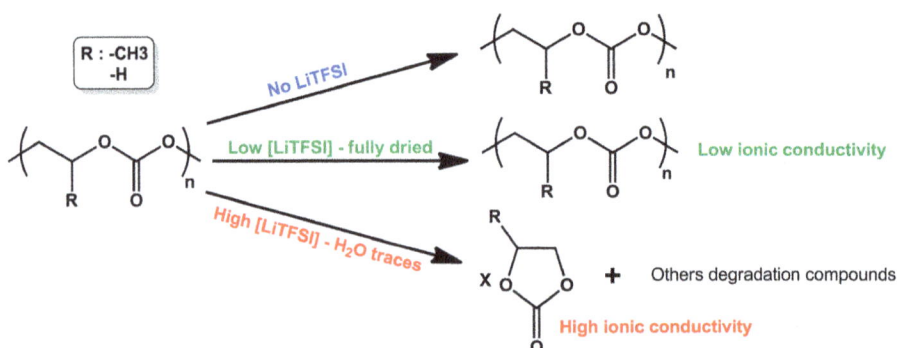

Fig. 5.18: Degradation reactions of PEC- and PPC-based SPEs. Adapted from [102], Copyright 2019, with permission from Elsevier.

With increased ring strain, six-membered cyclic carbonates can be readily polymerized with a high degree of control to form polycarbonates of the PTMC type. This type of materials has a long history of use in biomedical applications owing to the biodegradability of PTMC in combination with the synthetic flexibility of the monomer and corresponding polymer platform that allows for diverse functionalization. For SPEs, both PTMC and functionalized varieties have found use as host materials, as exemplified in Fig. 5.19a. The first use of this class in an SPE context utilized a triblock copolymer of poly(2,2-dimethyltrimethylene carbonate) and PEO [103]. In this system, the greater affinity of Li$^+$ for the PEO phase rendered the polycarbonate segments into passive anchoring units and the material functionally essentially a polyether electrolyte.

In order to utilize the ion-coordinating and -conducting properties of the polycarbonate, all ion-coordinating oxyethylene chains need to be eliminated. Similar to PEC, PTMC is largely amorphous, but has a lower T_g of around −15 °C at high molecular weights. Its excellent ion-solvating capabilities have been demonstrated in SPEs with a wide variety of both lithium and sodium salts [104–107]. Although the lower T_g of PTMC indicates notably higher molecular flexibility than for PEC, the ionic

a)

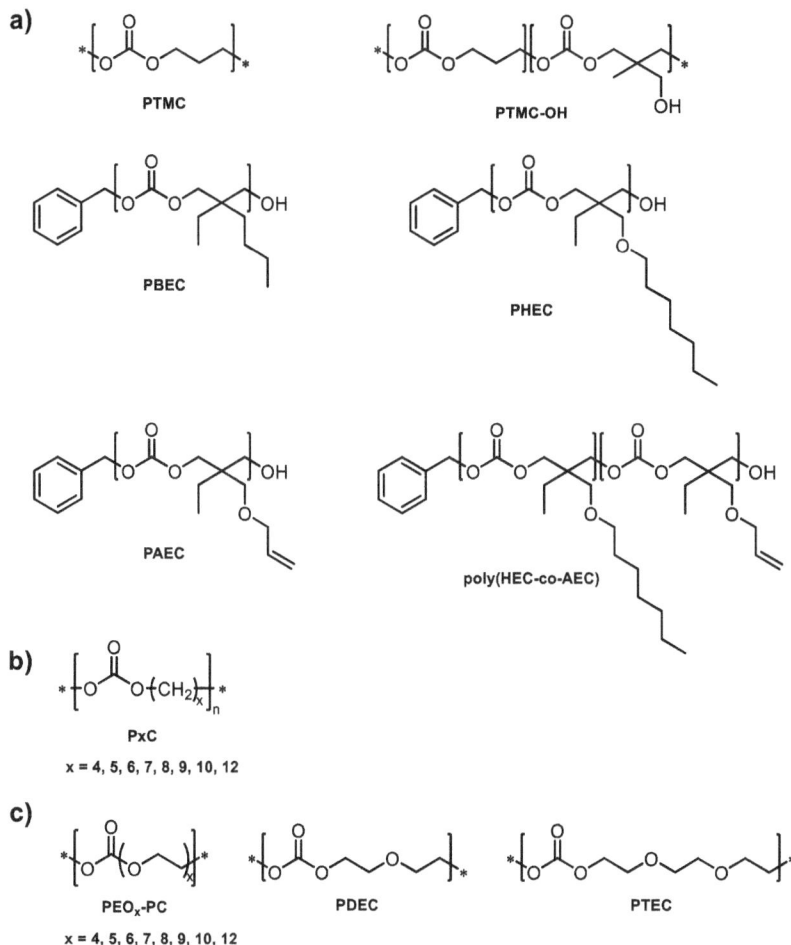

PTMC

PTMC-OH

PBEC

PHEC

PAEC

poly(HEC-co-AEC)

b)

PxC

x = 4, 5, 6, 7, 8, 9, 10, 12

c)

PEO$_x$-PC

PDEC

PTEC

x = 4, 5, 6, 7, 8, 9, 10, 12

Fig. 5.19: a) PTMC-type homo- and copolycarbonates; b) polycarbonates with ≥4 methylene groups in the repeating unit; and c) carbonate-linked oligoethers that have been investigated for use as host materials in SPEs.

conductivities remain lower in PTMC systems. This can be traced back to the lack of salt plasticization at moderate salt concentrations in PTMC. In contrast to PEC: LiTFSI, for example, the PTMC:LiTFSI system behaves much more similar to classic polyether systems, where the addition of salt leads to a stiffening effect which limits the ionic conductivity and leads to a clear conductivity maximum [105]. The behavior of PTMC with sodium salts largely mirrors that with lithium salts, but the effect of salt concentration on T_g is more pronounced. PISE-type conductivity has been demonstrated at high concentrations of NaFSI (Fig. 5.20). Under these conditions, the conductivity reaches as high as 5×10^{-5} S cm^{-1} at 25 °C for PTMC$_1$NaFSI [107].

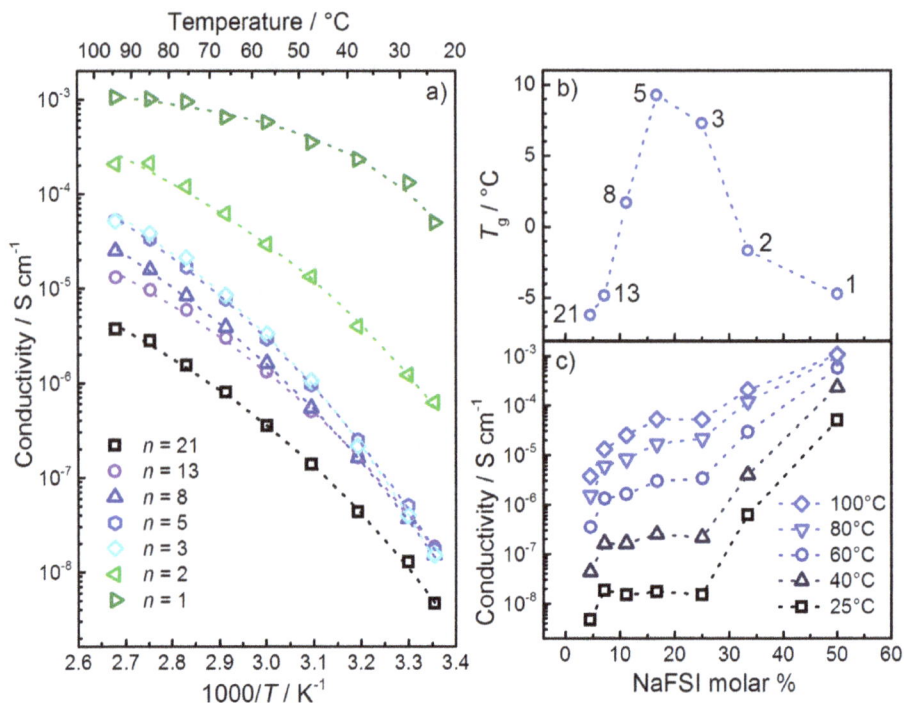

Fig. 5.20: Ionic conductivity behavior and thermal properties of PTMC$_n$NaFSI electrolytes. Reprinted with permission from [107]. Copyright 2019 American Chemical Society.

As already mentioned, the synthesis of PTMC from the cyclic monomer through ROP may be initiated from protic species. This includes water, which may naturally be found as a contaminant. Any water present during polymerization will therefore react and effectively be eliminated, leaving an anhydrous polymer. This is in sharp contrast with PEO, which is notable for containing traces of water that have proven difficult to completely eliminate. This has consequences for the electrochemical stability during battery cycling; whereas the interphase between PEO:LiTFSI and graphite contains large amounts of LiOH, no such traces can be seen in the equivalent PTMC system [26, 27].

The synthetic versatility of the six-membered cyclic carbonate monomer platform has been utilized to create functionalized and tailored materials through both monomer and post-polymerization functionalization strategies, with materials examples shown in Fig. 5.19a. These include polymers with flexible and plasticizing heptyl ether side chains (PHEC) and polymerizable allyl side groups (PAEC). The flexibility of these groups substantially lower the T_g down to −49 °C for pure PHEC and −48 °C for poly (HEC-*co*-AEC) [108]. This, however, does not translate into higher ionic conductivities for these materials (Fig. 5.21). This deficiency, which becomes exceedingly clear when also adjusting for the substantially lower glass transition temperatures (Fig. 5.21b),

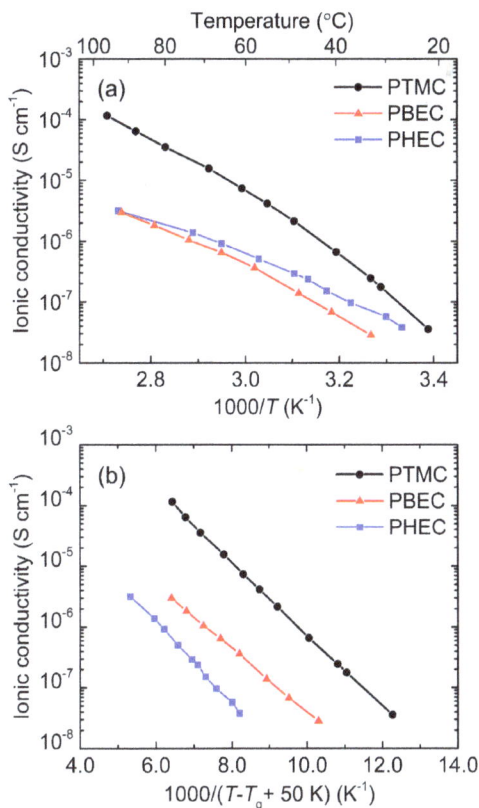

Fig. 5.21: (a) Total ionic conductivity of polycarbonate:LiTFSI electrolytes at a fixed salt concentration of [Li⁺]:[carbonate] = 0.08. (b) The same data presented on a shifted temperature scale to account for the differences in T_g between the systems. Reprinted from [57] under CC-BY 4.0 (http://creativecommons.org/licenses/by/4.0/).

can be traced to the presence of the bulky side groups that act to block the ion-coordinating polymer chains from coming into close proximity of each other to form suitable coordination environments. This results in severely reduced ion movement between coordination sites in host materials such as PHEC and PBEC compared to the bare-backbone PTMC [57]. These effects of the solvation site connectivity of the system are very similar to what has also been observed in polyethers with nonfunctional spacers [58] and highlights that a high molecular flexibility of the host material is not a guarantee for fast ion transport if the formation of suitable coordination environments is impeded.

With hydroxyl side groups in the polycarbonate structure, such as the copolymer PTMC-OH shown in Fig. 5.19, it is possible to introduce specific interactions between the SPE and nanostructured inorganic electrode materials to enable the formation of thin conformal coatings for, for example, 3D-microbatteries (see Chapter 4). An added effect of the hydrogen-bonding side groups is the possibility of interactions with both cations and anions. The result is an inverse dependence of T_g on salt concentration (decreasing T_g with increasing salt content; see Fig. 5.22) and very little variation in ionic conductivity with salt concentration as the interactions of the hydroxyl groups with the anion disrupt the hydrogen bonding between hydroxyl groups, thereby cancelling out the stiffening effect of these inter- and intramolecular hydrogen bond interactions [109].

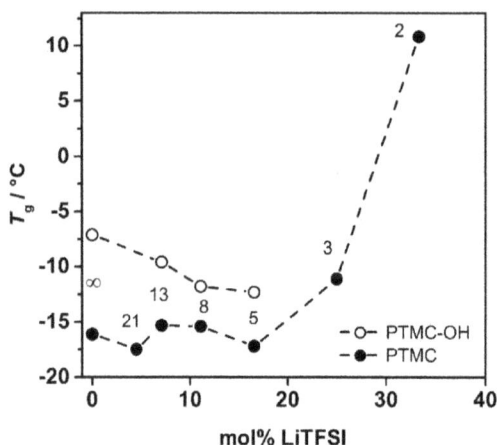

Fig. 5.22: Comparison of the variation of T_g with LiTFSI concentration for PTMC and PTMC-OH electrolytes. Data from [105, 109].

Using polycondensation, polycarbonates with more than two- or three-carbon blocks in the main chain can be obtained. Using the same synthesis method, polycarbonates containing oligo(ethylene oxide) blocks of arbitrary length can also be prepared. These types of materials are exemplified by the PxC and PEO$_x$-PC series of polymers (Fig. 5.19b–c) prepared by Meabe et al. [82, 91]. Several of these polymers – unlike PEC and PTMC – are semicrystalline, with melting points in the range of 47–63 °C for the PxC series and 29–52 °C for the PEO$_x$-PC series. The longer segments between the carbonate groups in the backbones of these polymers also lead to glass transition temperatures that are lower than for PEC or PTMC – down to –54 °C for P8C, and –55 °C for PEO$_{34}$-PC and PEO$_{45}$-PC. The oligoether/carbonate main chain combination of the PEO$_x$-PC series makes this a particularly interesting system for highlighting the differences in ion coordination between polyether and polycarbonate systems. Materials of the same type, based on di- or triethylene oxide (referred to as PDEC and PTEC, respectively, in Fig. 5.19c), have also been reported [110, 111]. With a structure in between that of polyethers and polycarbonates, cation transference numbers in between what is typical for these systems can be noted, for example, 0.39 for PTEC:LiTFSI.

Carbonate groups can also be incorporated in host materials for SPEs as (generally cyclic) side groups to the polymer backbone, either at the end of a spacer chain, directly attached to the main chain (vinyl carbonate-based) or by having the main chain pass directly through the cyclic carbonate moiety (vinylene carbonate-based). Examples of such structures are shown in Fig. 5.23.

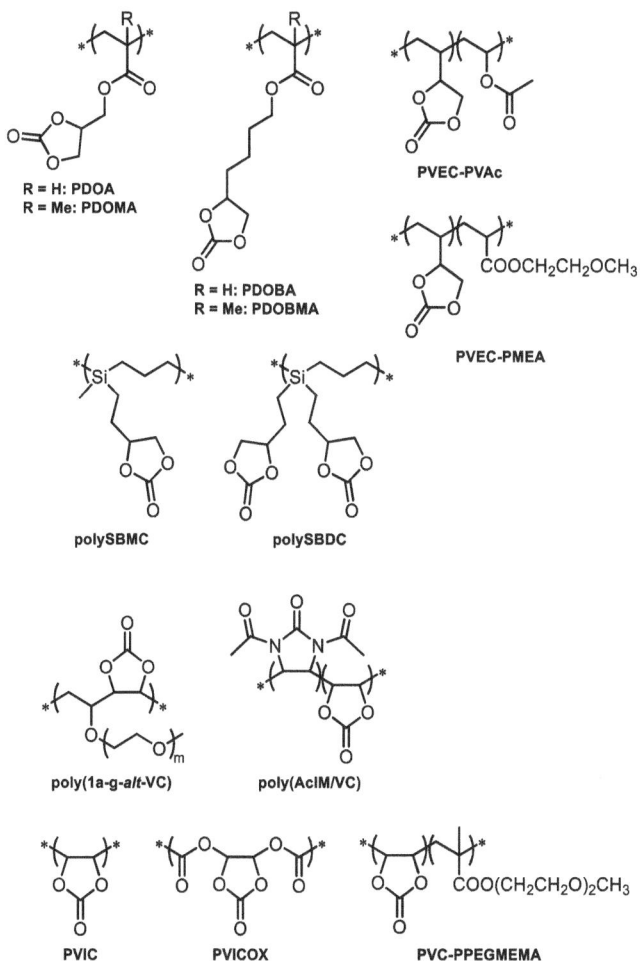

Fig. 5.23: Polycarbonates with the carbonate moieties as side groups.

In systems where the carbonate group is directly and firmly anchored to the polymer backbone, much more rigid systems are obtained than when a flexible spacer is used. In such systems, decoupled "superionic" conductivity has been noted, that is, a rate of ionic motion that is far higher than the rate of structural relaxation. Examples of host materials where such behavior has been noted are PVIC, PVICOX, PVEC-PVAc, PVEC-PMEA, PVC-PPEGMEMA and poly(AcIM/VIC) (Fig. 5.23). For PVIC:LiCF$_3$SO$_3$, a conductivity on the order of 10^{-7} S cm^{-1} at 40 °C has been reported at high salt concentrations, while at the same time being so rigid that it is difficult to even detect a T_g at all, and even higher conductivities are seen for PVICOX:LiCF$_3$SO$_3$ at similar salt concentrations [112]. For PVEC-PVAc, PVEC-PMEA and PVC-PPEGMEMA electrolytes with LiTFSI, decoupled conductivity behavior is seen already at low salt concentrations [113]. Although the phenomenon of ionic conductivity in rigid matrices is highly attractive, the high glass transition temperatures (in combination with low salt concentrations for the latter systems), results in low ionic conductivities compared to traditional low-T_g systems, but still signal the possibility of going beyond the confines and limitations of coupled ion transport with the use of polycarbonate-based host materials.

5.2.3 Polyesters

Carboxylate esters are in many ways similar to the carbonate esters already discussed in Section 5.2.2. While ester solvents such as ethyl acetate and γ-butyrolactone have found use in solvent mixtures for liquid electrolytes to, for example, improve the low-temperature performance, organic carbonates are much more ubiquitous for liquid-electrolyte use. When it comes to their polymeric analogues, *polyesters* synthesized through either polycondensation or ROP may also interact with and solvate metal cations by means of their carbonyl functionalities. Similar to polycarbonates, the alcohol-residue oxygen in the main chain is not observed to be involved in ion coordination to any greater extent [89]. Compared to polycarbonates, recent data suggest stronger Li$^+$ coordination for polyesters, but weaker compared to PEO [25]. This translates into polyester T_+ values in between those for polycarbonates and polyethers.

While largely remaining in the background relative to PEO, a wide range of main-chain polyesters were indeed explored for SPE use already during the 1980s. The diversity of polyester hosts for SPEs is illustrated in Tab. 5.3. Most of these host polymers are in many aspects very similar, with glass transition temperatures well below room temperature and semicrystallinity with melting points in the relatively narrow range of 55–79 °C. These properties are also in many cases quite similar to the thermal properties of PEO. Much like for PEO, the semicrystallinity of these host materials is also to a large degree retained in the respective SPEs, restricting ion transport. Unlike PEO, there have not been any reports of crystalline electrolyte complexes in polyester systems, with the possible exception of poly(β-propiolactone):LiClO$_4$ [114]. Similar to PEO-based electrolytes, the T_g increases with salt concentration while the T_m and degree of crystallinity decrease accordingly. Due to the semicrystallinity, the ionic conductivities may be somewhat limited at room temperature, but values as high as 1.06×10^{-5} S cm^{-1} at 30 °C for poly(ethylene adipate):LiClO$_4$, 1.6×10^{-6} S cm^{-1} at 25 °C for poly(ethylene malonate):LiCF$_3$SO$_3$ and 1.2×10^{-6} S cm^{-1} at room temperature for poly(ε-caprolactone):LiClO$_4$ have been reported.

Of these polyesters, poly(ε-caprolactone) (PCL) is the most extensively investigated material. With thermal properties that closely resemble those of PEO, but with weaker ion complexation, PCL can deliver similar ionic conductivity as PEO combined with a

Tab. 5.3: Compilation of structures and properties of polyesters used as host materials for solid polymer electrolytes.

Structure	Repeating unit	T_g (°C)	T_m (°C)	References
	Ethylene adipate	−49	58–62	[115, 116]
	1,4-Butylene adipate	−60	56–60	[115, 116]
	1,6-Hexamethylene adipate	−62	55–65	[115, 116]
	Ethylene succinate	−6–(−1)	100/103	[114, 117, 118]
	Ethylene malonate	−15	n.c.[a]	[119]
	Ethylene sebacate	−42	79	[117]
	β-Propiolactone	−21	74/77	[114]
	ε-Caprolactone	−65–(−64)	60	[115, 116, 120–122]

[a]n.c. represents not crystalline.

notably higher T_+. Similar to PEO, crystallinity is also a major limit to performance below the melting point. This can to some extent be mitigated by increasing the salt concentration or adding inorganic nanoparticles, as shown in Fig. 5.24. Another option is to disrupt the crystallinity by copolymerization. The addition of 20 mol% trimethylene carbonate repeating units to form a random copolymer almost completely eliminates crystallinity in the pure host material, with any remaining crystallinity eliminated by the addition of salt. The inclusion of the carbonate units further serves to reduce the overall ion binding strength for optimal cation dynamics. This material can support an ionic conductivity of up to 4.1×10^{-5} S cm^{-1} at 25 °C with LiTFSI and 1.28×10^{-5} S cm^{-1} at 25 °C with NaFSI salt [120, 123]. The T_+ appears to be lower for Na$^+$ than for Li$^+$, although it should be noted that it is far more difficult to get accurate measurements of transference numbers in sodium systems as sodium metal is much more reactive than lithium metal. Similar to polycarbonate systems, the effect of salt concentration on T_g is much more pronounced in sodium than in lithium systems, and the maximum conductivity can therefore be found at relatively low salt concentrations.

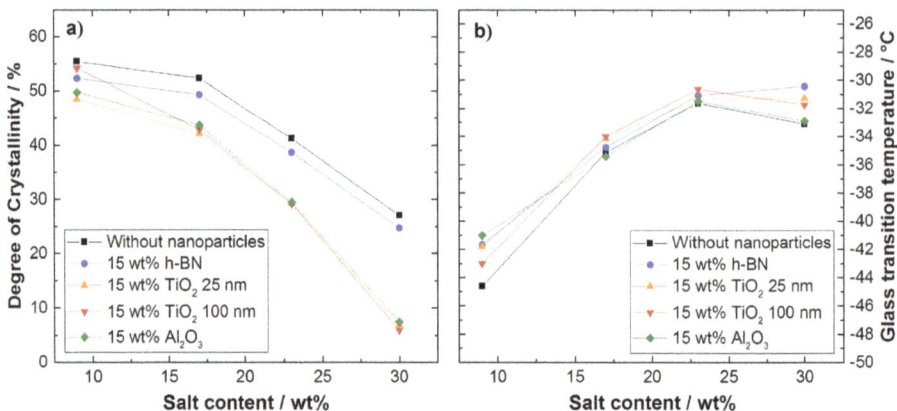

Fig. 5.24: a) Degree of crystallinity and b) glass transition temperature as a function of salt content for PCL:LiTFSI electrolytes with or without 15 wt% inorganic nanoparticles. Reprinted from [124], Copyright 2019, with permission from Elsevier.

Instead of utilizing a polyester main chain, cation-coordinating ester moieties may also be located in pendant side groups. There is indeed a great variety of such materials available based primarily on the acrylate and methacrylate materials platforms. Despite this ready availability, this class of polymers has seen very limited use for SPEs, although frequent application for gel polymer electrolytes can be noted. Solvent-free electrolytes based on poly(methyl methacrylate), poly(butyl acrylate) and copolymers of these with acrylonitrile have nevertheless been reported [125–127].

5.2.4 Polyketones

Aliphatic polyketones is a far less developed class of polymers than polycarbonates or polyesters, with less straightforward procedures required for their synthesis. As a consequence of this lack of available host polymers, polyketones have only recently begun to be investigated for SPE use. Poly(3,3-dimethylpentane-2,4-dione) is a semicrystalline polyketone with a relatively high T_g of slightly more than 40 °C and has a structure with two ketone groups in each repeating unit, allowing for chelating coordination structures to form with metal cations [128]. This property leads to an anomalous shift of the carbonyl stretch IR vibration to higher wavenumbers on coordination due to conformal changes. When LiTFSI salt is added, the T_g is decreased down to –9 °C (a similar range as for PTMC-based electrolytes, as discussed in Section 5.2.2) and the crystallinity diminishes, leaving electrolytes with 40 wt% LiTFSI fully amorphous. The conductivity in this material is limited by the T_g to the order of 10^{-7} S cm^{-1} at 25 °C. The T_+ of 0.7 for poly(3,3-dimethylpentane-2,4-dione) with 25 wt% LiTFSI is in a similar range as for other carbonyl-coordinating host polymers, indicating similarly weak cation coordination in polyketone systems.

5.2.5 Application of carbonyl-coordinating polymers in batteries

Reflecting the diversity of available host materials, a wide variety of carbonyl-coordinating polymers have been successfully implemented in SPEs in solid-state battery prototypes. A fair share of these are also capable of delivering high performance at ambient temperature, supported by the notably high T_+ and low crystallinity generally observed for these materials. At a high salt concentration, PEC:LiFSI functions well in Li ǁ LFP batteries, delivering high performance at down to 30 °C [68]. High-performance battery cell data has also been reported for PPC-based electrolytes [98], including cycling at room temperature of a potassium-based system with an organic 3,4,9,10-perylene-tetracarboxylic dianhydride (PTCDA) cathode and a PPC:KFSI electrolyte supported by a cellulose nonwoven separator membrane [100]. None of these electrolyte systems has sufficient intrinsic mechanical stability, and consequently need to be combined with separator membranes for reliable implementation in batteries.

The lower conductivity of PTMC-based SPEs necessitate an operating temperature of 60 °C for reasonable performance in Li ǁ LFP cells [105]. A peculiar feature of PTMC-based cells is a gradual evolution of the delivered capacity with time – even when the electrolyte is cast directly onto the porous LFP cathode – caused by an insufficient electrolyte–electrode interface [129], as also discussed in Chapter 4. At elevated temperatures, this interface slowly develops over time until the full capacity of the cathode can be utilized. In sodium systems, this effect is much less pronounced. Half-cells with a PTMC:NaFSI electrolyte can be cycled with high stability at down to 40 °C. As the salt concentration is increased toward the PISE regime, the conductivity increases dramatically, but at the same time the cycling stability deteriorates, likely because of limited stability of the salt toward sodium metal [107].

The polyether/polycarbonate hybrid PTEC has been implemented in half-cells versus both the traditional LFP and the higher-voltage material $LiFe_{0.2}Mn_{0.8}PO_4$ with good stability at down to room temperature (Fig. 5.25). Because of low molecular weights, the limited mechanical properties of the material demanded the support of cellulose nonwoven substrates for mechanical stabilization [110]. Another polycarbonate that has been used together with higher-voltage cathode materials is PVIC, which has been successfully used in Li ǁ $LiCoO_2$ cells [130]. Using a PVIC:LiDFOB electrolyte prepared through *in situ* bulk polymerization, reliable cycling stability at 50 °C and up to C/2 was obtained.

When it comes to polyesters, materials based on PCL have seen the most success. PCL:LiTFSI electrolytes with 30 wt% salt can be cycled in Li ǁ LFP cells at reasonably low temperatures, delivering gradually increasing capacities with time at 5 µA cm^{-2} and 30 °C despite a fair degree of crystallinity in the material [124]. With added trimethylene carbonate units in the PCL main chain to both reduce the crystallinity and boost the Li$^+$ transport properties, stable and reliable room temperature performance can be obtained (Fig. 5.26) [120]. The same polymer has also been successfully applied

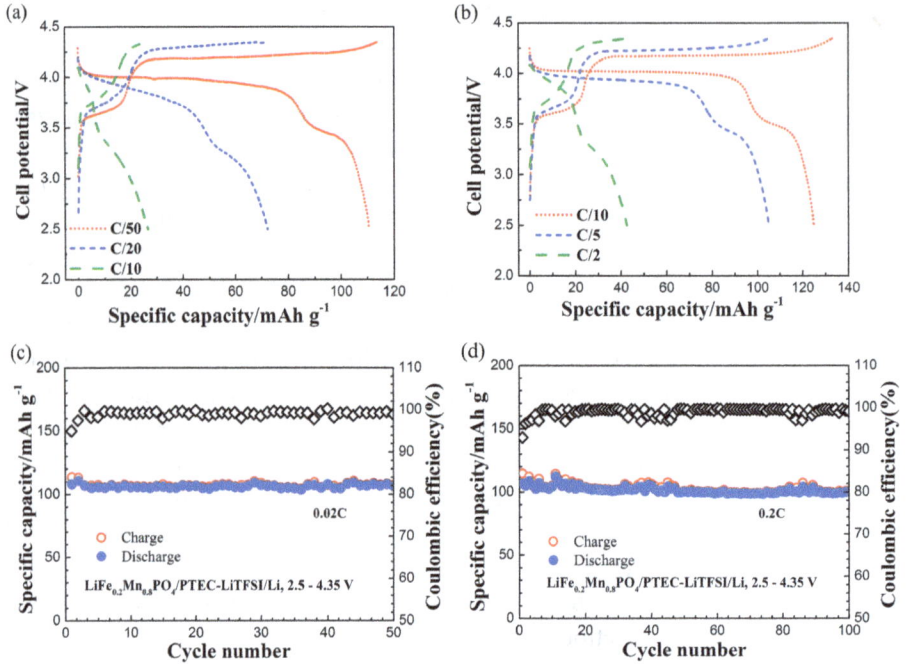

Fig. 5.25: Cycling performance of Li | PTEC:LiTFSI | LiFe$_{0.2}$Mn$_{0.8}$PO$_4$ half-cells at 25 °C (left) and 55 °C (right). Reprinted from [110], Copyright 2016, with permission from Elsevier.

to full-cell sodium batteries operational both at 40 °C and room temperature (Fig. 5.27) [123].

In addition to traditional half- and full-cell battery architectures, polycarbonate and polyester electrolytes have also been applied in 3D-microbattery architectures, where 3D-structured electrodes maximize the surface area of the electroactive material for efficient footprint usage. To enable reliable operation of such batteries, electrolytes that can be applied as thin, conformal coatings on the electrode surface are necessary. To this end, cycling of cells consisting of Cu$_2$O-covered Cu nanopillars with a poly(ε-caprolactone-*co*-trimethylene carbonate): LiTFSI electrolytes versus Li metal has been demonstrated, as shown in Fig. 5.28 [131].

5.3 Polynitriles

Besides O atoms seen for polyethers and carbonyl-based SPEs, polymers with N-containing functional groups constitute another family of host materials with potential applications as SPEs. In particular, molecules with the nitrile (cyano) group, such as acetonitrile, succinonitrile and adiponitrile, have been widely used in liquid electrolyte formulations [132, 133]. This group has also been found in ion-

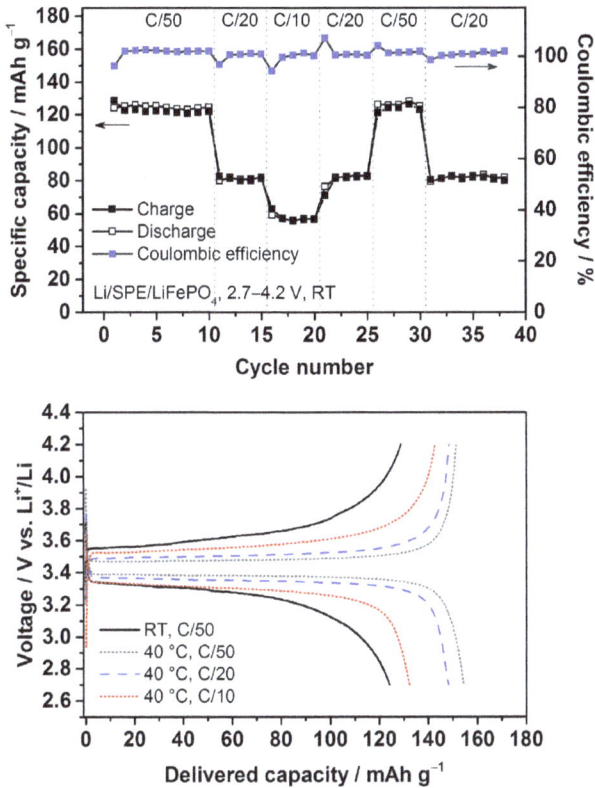

Fig. 5.26: Cycling performance of Li | poly(ε-caprolactone-*co*-trimethylene carbonate):LiTFSI | LFP half-cells. C/10 is equivalent to 0.027 mA cm^{-2}. Reprinted from [120], Copyright 2015, with permission from Elsevier.

conducting molecular plastic crystals [134] and as additives and plasticizers in different polymer electrolyte systems [135, 136]. With a proven ability of the nitrile group to coordinate to Li$^+$, it has also been used as SPE, polyacrylonitrile (PAN, structure shown in Fig. 5.29) being the most common polymer host bearing this functionality.

PAN is commercially synthesized by free-radical polymerization of acrylonitrile. It is a semicrystalline polymer with a T_g of 80 °C, it has a high dielectric constant, strong coordination ability and high oxidation potential (4.5 V vs Li$^+$/Li) [137, 138]. Considering the conventional mode of transport in SPE where cations are coupled to the segmental motion of the polymer host, at a first glance at the physical properties of this polymer (semicrystallinity and high T_g), poor cation movement would be expected. Despite this, PAN has been reported as a host material with different salts, such as LiClO$_4$, LiCF$_3$SO$_3$, LiBF$_4$, LiTFSI, LiFSI and LiBOB [138]. In most cases, the salt has a plasticizing effect on the polymer, lowering the T_g of the system. In

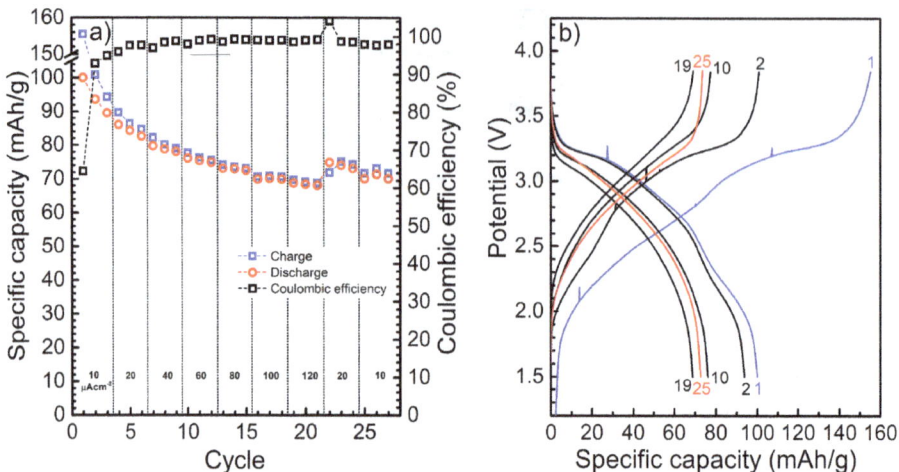

Fig. 5.27: Cycling performance of a hard carbon | poly(ε-caprolactone-co-trimethylene carbonate): NaFSI | $Na_{2-x}Fe(Fe(CN))_6$ full cell cycled at 40 °C. Reprinted from [123], Copyright 2019, with permission from Elsevier.

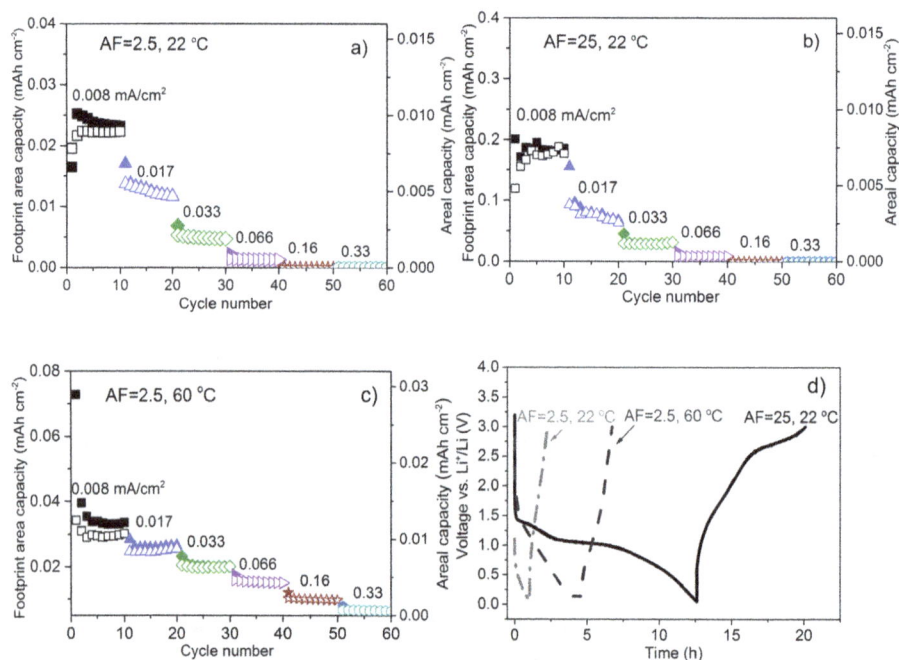

Fig. 5.28: Cycling of Li | poly(ε-caprolactone-co-trimethylene carbonate):LiTFSI | Cu_2O 3D-structured cells with varying area gain factors (AF) and temperatures. Reprinted with permission from [131]. Copyright 2017 American Chemical Society.

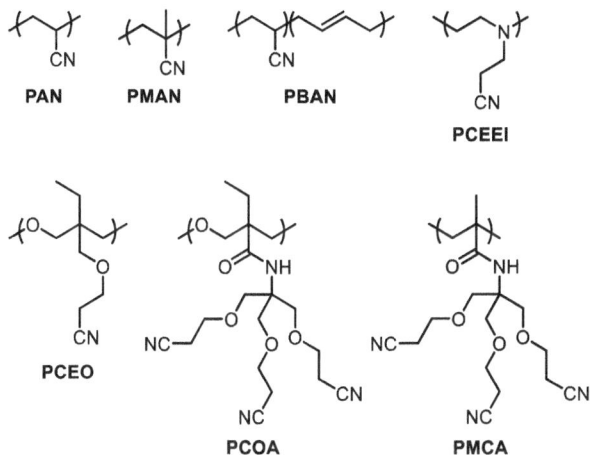

Fig. 5.29: Chemical structure of SPEs containing nitrile functionalities.

addition, by increasing the salt concentration, sometimes even up to the PISE regime, the T_g is further reduced and the conductivity enhanced [125].

One of the main challenges of using PAN is its low solubility in low-boiling-point solvents preferred for solvent casting the SPE. Typical solvents that dissolve PAN are N,N-dimethylformamide (DMF) and dimethylsulfoxide (DMSO) with high boiling points and that are usually very hard to completely remove. This means that significant levels of solvent residues – as much as 10 wt% [139] – are often found in the resulting SPEs even after prolonged and thorough drying conditions. The remaining solvent residues can interact with Li^+, as their donor number is higher than that of the nitrile. In addition, it has been observed that the amount of DMSO residues is only dependent on the amount of Li salt and not the polymer, indicating preferential coordination of Li^+ to DMSO and marginal – if any – interaction between $LiBF_4$ and PAN. This, consequently, drastically changes the ion conduction mechanism. Regardless of the amount of salt present in the PAN-based SPE (salt-rich or polymer-rich), the PAN does not participate in the ion transport; instead the only mobile species are $Li(DMSO)_n$ complexes and not the small fraction of Li^+ coordinated to PAN [140]. Similar behavior has been found when using ethylene carbonate [141] or DMF [142] as plasticizers. These examples indicate that the solvent residues are responsible for Li^+ mobility. Furthermore, it also implies that the lower the coordination of Li^+ to the polymer host, the higher the ionic conductivity.

In order to ensure a truly solvent-free SPE with PAN as the polymer host, mixtures of PAN with Li salts can be hot-pressed. With this preparation method, the ionic conductivities are lower compared to systems containing solvent residues. Nevertheless, increasing the salt concentration up to the PISE regime (above 60 wt% salt) decreases the T_g to 50–65 °C and increases the ionic conductivity 5 orders of magnitude [143]. These systems with high salt concentration show high ionic conductivities even at or below T_g (Fig. 5.30), indicating that the conduction mechanism is decoupled from the polymer segmental motion [144]. The more efficient ionic transport mechanism observed for the high salt concentration is associated with a higher degree of ionic aggregation and reflects the connectivity effects when salt-rich clustered domains come into contact, thereby creating connected pathways for ion transport [145]. At this percolation threshold, the salt becomes a continuous phase and the polymer merely acts as a plasticizer to inhibit crystallization of the salt [144].

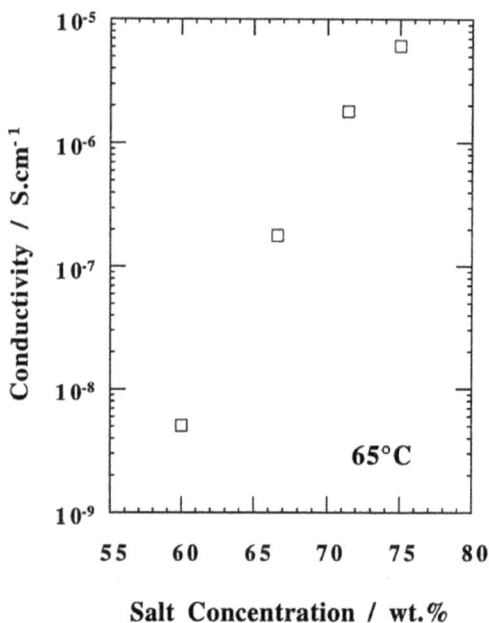

Fig. 5.30: Conductivity dependence on salt concentration for PAN:LiCF$_3$SO$_3$ electrolytes at 65 °C (near the expected T_g). Reprinted with permission from [144]. © 2000 John Wiley & Sons.

Another approach to employ PAN as polymer host is to decrease its molecular weight to the oligomer range (between 1,000 and 2,000 g mol^{-1}). This allows the use of low-boiling-point solvents such as acetonitrile or dioxolane, decreases the T_g from 70 °C for the pure oligomer to 14 °C with 30 wt% LiFSI and increases the ionic conductivity [146].

Due to the low ionic conductivity observed for the PAN-based SPEs, only a few reports have demonstrated their utilization in battery devices. One example is a PAN:LiClO$_4$ electrolyte containing silica aerogel powder (SAP) cast from a solution of DMF onto LFP. The capacity was stable for 20 cycles at C/2 but the Coulombic efficiency remained rather low, around 94% (Fig. 5.31) [147]. Despite the poor mechanical stability provided by the PAN oligomer, its decent ionic conductivity made it functional as SPE in NMC622 || Li cells. Although the cell was only cycled three times, it featured a high capacity of 150 mAh g^{-1} at C/10 [146].

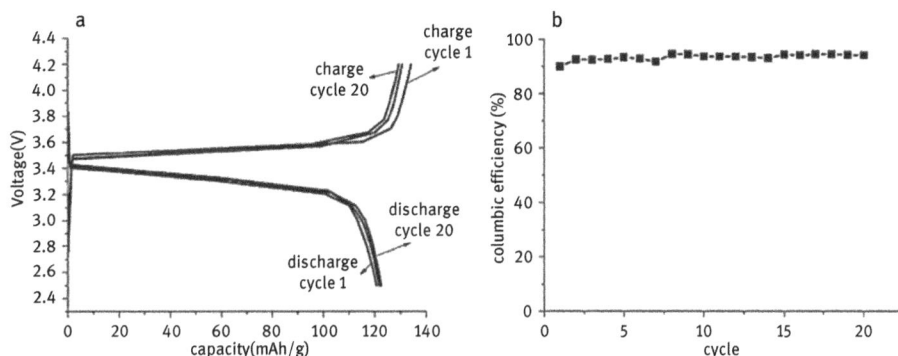

Fig. 5.31: Cycling performance of a Li | PAN:LiClO$_4$ + SAP | LFP battery at C/2 and room temperature. Adapted from [147], Copyright 2010, with permission from Elsevier.

5.3.1 Polyacrylonitrile derivatives and copolymers

With the aim to overcome the issue with solvent residues in PAN-based SPEs, PAN derivatives soluble in low-boiling-point solvents have been developed. Poly(methacrylonitrile) (PMAN, structure shown in Fig. 5.29), for example, is soluble in acetone but it is more rigid than PAN with a T_g around 120 °C. However, upon addition of LiTFSI up to the PISE regime, the material softens and reaches a reasonable ionic conductivity, 10^{-4} S cm^{-1}, at 90 °C with an Arrhenius-type temperature dependence [148].

A different strategy to increase the solubility in volatile organic solvents is through copolymerization of acrylonitrile with other comonomers. Poly(acrylonitrile-co-butadiene) (PBAN, structure shown in Fig. 5.29) can be used as polymer host with LiAsF$_6$, LiCF$_3$SO$_3$ and LiClO$_4$ [149, 150]. The butadiene comonomer avoids the crystallization of PAN, rendering it an amorphous elastomer with high chain flexibility ($T_g = -42$ °C) and solubility in methyl ethyl ketone. These SPEs show a VFT-type temperature dependence with a maximum ionic conductivity at 11 mol% LiClO$_4$ for the PBAN electrolyte. Similarly, amorphous polymers can be obtained when copolymerizing acrylonitrile, itaconic acid and methacrylic acid mixed with LiClO$_4$ salt [150]. A high ionic conductivity was achieved close to the limit of salt solvation, suggested to be due to the formation of an "infinite cluster" when all the separate single clusters come into contact, promoting fast cationic transport. The presence of residual solvent in these systems, either DMF or ethyl methyl ketone, results in lower ionic conductivity, in contrast to what has been described previously for PAN-based electrolytes. This behavior was explained by the tightly linked Li$^+$ to the carbonyl group of the solvent molecules that decreases the cation transport and confirmed with a lower t_+ value [18].

A similar approach can be carried out with butyl acrylate as comonomer with acrylonitrile, yielding copolymers soluble in acetonitrile [125, 126]. The ionic conductivity and mechanical properties are highly dependent on the salt identity. Salts with strong interactions with the polymer host, such as LiTFSI and LiI, act as plasticizers rendering amorphous matrices with a T_g of -22 and -6.5 °C, respectively. In contrast to PEO-based electrolytes, no physical cross-linking and no stiffness of the chain occur, even at low salt concentrations, but the flexibility instead increased [125, 126]. Despite the amorphous nature and low T_g values with LiTFSI and LiI, these SPEs show low ionic conductivity. Incorporating salts with weaker interactions with the polymer host, such as LiCF$_3$SO$_3$ and LiAlCl$_4$, results in more mobility of the ions and thereby SPEs with higher ionic conductivity [125]. The weaker interactions of these salts with the polymer host are indicated by the larger amounts of nitrile groups that are left uncomplexed. This is consistent with what has been observed for PAN-based systems – less coordination of Li$^+$ ions to the nitrile groups is equivalent to higher ionic conductivity.

For highly concentrated samples of PBAN:LiAsF$_6$, the T_g and ionic conductivity have a broad distribution of experimental values that suggests the presence of a metastable state of the system [151, 152]. This complicated feature can be attributed to specific structural transformations of the SPE, changing from a rubbery state for the just-prepared SPE to a brittle state after long-term storage [152]. It has also been found that the T_g is defined by the preparation conditions (casting solution concentration, solvent evaporation rate, etc.) and the thermal pre-history, rather than by salt concentration [151]. This behavior has also been observed for poly(acrylonitrile-co-butylacrylate):LiTFSI in the PISE regime. Upon prolonged storage of the SPEs, the glass transition temperature increases, the ionic conductivity decreases – as shown in Fig. 5.32 – and precipitation of salt can be observed at the nanometer length scale for samples containing more than 84 wt% of salt. The aging effects can also be related to the loss of structure and continuity of the conductivity pathways [127].

5.3.2 Other nitrile-functional polymers

Besides using acrylonitrile as the primary monomer feedstock to prepare PAN-based electrolytes, it can also be used to functionalize other monomers [153] or polymers [63–65] through the Michael addition reaction, forming propanenitrile side groups. This strategy can be carried out to modify polyethylenimine (PEI) (another type of polymer host described in Section 5.4) forming poly((N-2-cyanoethyl)ethylenimine) (PCEEI, structure in Fig. 5.29). This polymer host structure disrupts the crystallinity typical of PEI, decreases the T_g (-36 °C) compared to that of PAN and is soluble in acetonitrile. Despite all these advantages, the ionic conductivity is still rather low when doped with LiCF$_3$SO$_3$, on the order of 10^{-8} S cm^{-1} at room temperature [153].

Another family of nitrile-functional polymers are hybrids of polyethers and polynitriles forming cyano-functional polyoxetanes and polymethacrylamides (PCEO, PCOA and PMCA, structures are shown in Fig. 5.29). In these SPEs, the cations coordinate to both ether oxygens and nitrile groups [63–65]. Addition of lithium salts,

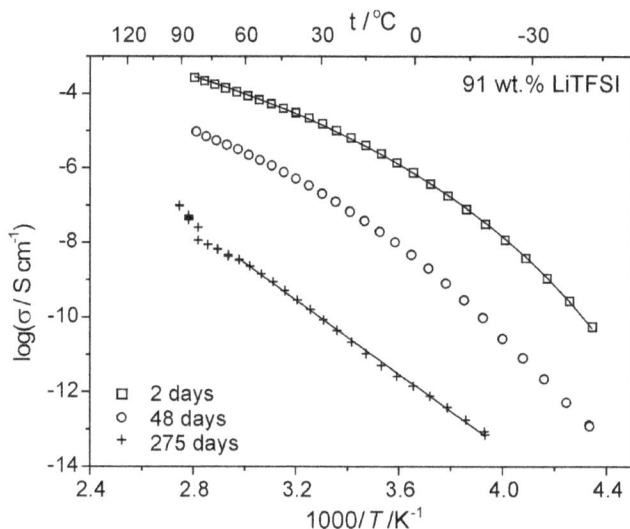

Fig. 5.32: Aging effects on the ionic conductivity of PBAN:LiTFSI electrolytes with a salt concentration of 65%. Reprinted from [127], Copyright 2015, with permission from Elsevier.

either LiTFSI or LiBF$_4$, yields amorphous polymers (T_g – 14 °C for PCOA and –21 °C for PCEO) with ionic conductivities that show VFT-type temperature behavior [63, 64]. The low molecular weight obtained for PCOA (4,100 g mol^{-1}) requires the addition of, for example, PVdF-HFP as a mechanical support, and the ionic conductivity of this system with LiBF$_4$ is higher (2×10^{-4} S cm^{-1}) [64] than the analogous PCEO:LiTFSI (2.5×10^{-5} S cm^{-1}) [63]. The coordination strength of these two SPEs has been studied by IR and the results suggest that the number of coordinated lithium ions in the PCOA electrolyte is larger than in the PCEO films [64]. This is contrary to what is observed for PAN-based electrolytes where lower coordination of the cation to the nitrile group results in higher ionic conductivity. To further improve the mechanical properties, the polymer backbone can be changed from polyether in PCOA to methacrylamide in PMCA. Addition of LiTFSI to this polymer host yields a flexible and self-standing SPE but with lower ionic conductivity (10^{-6} S cm^{-1}) than PCOA and lower oxidation stability (3.1 V vs Li$^+$/Li) than PAN [65].

Overall, nitrile- or cyano-based polymers are able to dissolve large amounts of salt, they have a strong coordination ability and possess high anodic oxidation potential. All these advantages make them suitable polymer hosts for SPEs. However, there are still many challenges that limit their use in battery devices, the main ones being the presence of solvent residues that completely changes the ion transport mechanism and the low ionic conductivity without these residues. New designs, further modifications and understanding of the nitrile-based SPEs are still required to overcome these challenges and allow the application of polynitriles in high-energy-density all-solid-state batteries. Perhaps it is not as bulk SPEs that polynitriles have their most prosperous future in battery applications, but as surface coatings and other types of components in high-voltage cathodes.

5.4 Polyamines

A direct analogue of PEO where the oxygen is replaced with nitrogen (NH) is polyethylenimine (PEI or LPEI when the polymer has a linear structure, Fig. 5.33) [154]. PEI is able to dissolve alkali metal salts as the cations coordinate to the nitrogen of PEI, similarly to the oxygens in PEO; however, in PEI–PEO–PEI block copolymers, Li$^+$ is preferentially coordinated by the etheric oxygen of PEO than by the nitrogen of PEI [155]. An interesting feature of PEI is the ability of the NH group to form hydrogen bonds with the anions, which is beneficial for polymer hosts in SPEs. However, the hydrogen bonding ability also renders a highly crystalline pristine polymer (up to 80%) [156] with a melting point at 60 °C for high-molecular-weight LPEI [157]. The excessive crystallinity of LPEI makes the determination of the glass transition temperature very difficult. Semicrystalline LPEI has a T_g of −23.5 °C [158], although extrapolation of data from more amorphous systems indicate a T_g closer to −40 °C [156]. LPEI is synthesized through cationic ROP of 2-oxazoline followed by alkaline hydrolysis to remove the acyl group [159]. A more conventional synthesis route of PEI is cationic polymerization of aziridine; however, this method yields a highly branched structure (BPEI, Fig. 5.33) [160]. While LPEI is highly crystalline and only contains secondary amino groups, BPEI is amorphous (T_g is −47 °C) [161] and contains additional primary and tertiary amino groups due to its branched structure.

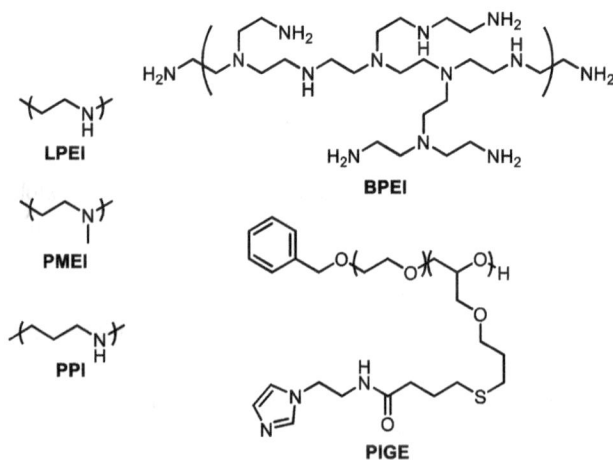

Fig. 5.33: Chemical structure of SPEs containing amino functionalities.

Upon addition of alkali metal salts, the crystallinity of PEI is suppressed due to the strong interactions between both components. Small amounts of NaI in LPEI decrease the crystallinity of the mixture and crystallization is completely suppressed at a molar ratio of 0.15 NaI/LPEI [162]. LPEI is also able to dissolve many lithium salts (LiCF$_3$SO$_3$, LiCl, LiBr, LiI, LiSCN, LiClO$_4$ and LiBF$_4$) [156]. The crystallinity of the polymer decreases upon the addition of the salts, although to a different degree depending on the anion and their lattice energies. LiCF$_3$SO$_3$ with a low lattice energy is the most readily dissolved and has the greatest effect on crystallinity. In addition, the glass transition temperature increases upon addition of salt, similar to PEO-based SPEs. At concentrations above 10 mol% the mixture remains amorphous with an ionic conductivity of 10^{-8} S cm^{-1} at room temperature and 10^{-3} S cm^{-1} at 150 °C [156]. The decrease in crystallinity upon addition of salt has also been studied with FTIR, showing that the hydrogen bonding is disrupted with increasing salt concentration or temperature [163, 164]. While in the LPEI:LiSbF$_6$ system, the primary interaction of the anion is with the NH group of the polymer, the same behavior is not observed in the LPEI:LiCF$_3$SO$_3$ system [164].

Besides being a host material, LPEI can also be used to dope PEO:LiClO$_4$ electrolytes and suppress the crystallization of PEO, leading to a 100-fold increase in ionic conductivity [165, 166]. These SPEs were cast from a methanol solution and solvent residues were found in the samples leading to an increase in ionic conductivity of about half an order of magnitude [166]. As methanol is a commonly used solvent for PEI-based SPEs, it might also be present in other reported systems without being noticed.

Pure branched PEI is amorphous and addition of LiCF$_3$SO$_3$ salt increases the T_g until a semicrystalline phase is formed at high salt concentrations of N/Li = 4 with a T_m at 49 °C. The conductivity of samples with low salt concentration follows a VFT-type temperature dependence and the maximum value is observed at N/Li = 20. Cation–nitrogen coordination and hydrogen bonding between N–H and anions contribute to a weakening of the N–H bond, clearly observed with IR spectroscopy. Furthermore, ion pairing can be detected at higher salt concentrations [161]. When adding LiTFSI to BPEI, the matrix initially softens and the ionic conductivity increases until N/Li = 100 where T_g and ionic conductivity show a minimum and maximum, respectively (Fig. 5.34). Higher salt concentrations lead to salt bridging and a rapid increase in T_g [167]. Although an Arrhenius dependence of the conductivity versus temperature has been observed for this system, this behavior is unexpected and it is still not clear why or what underlying ion transport mechanism is responsible [168].

Another strategy to suppress the crystallinity of PEI is to synthesize its analog poly(N-methylethylenimine) (PMEI, Fig. 5.33) [154] through methylation of the nitrogen in PEI [157]. PMEI is unable to form hydrogen bonds and is completely amorphous (T_g is −82 °C). Incorporation of LiClO$_4$ or LiCF$_3$SO$_3$ to PEI or PMEI increases the T_g of the system as the Li$^+$ ions form physical cross-links. At high salt concentration the ionic conductivity is controlled by the T_g (coupled to segmental motions)

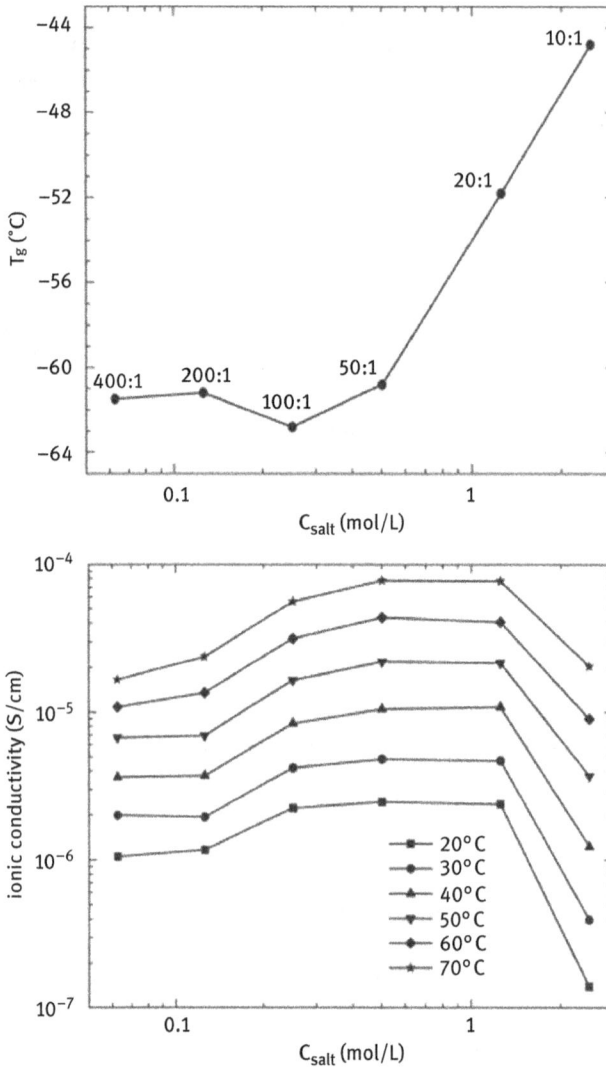

Fig. 5.34: Variation of T_g (top) and ionic conductivity (bottom) in BPEI:LiTFSI electrolytes showing both slight plasticization at low salt concentrations and stiffening due to salt bridges, with a concomitant drop in conductivity at high concentrations. Adapted from [167], Copyright 2010, with permission from Elsevier.

irrespective of the polymer host or anion, whereas at low salt concentration the PEI system is more conductive than PMEI for both salts, as the latter is a weaker salt dissociating matrix (Fig. 5.35) [154].

A similar structure to PEI can be obtained by adding another methylene group to the repeating unit of the polymer, thereby giving polypropylenimine (PPI, Fig. 5.33). Similar to PEI, PPI is highly crystalline with T_m of 60.5 °C. However, while PEI is highly

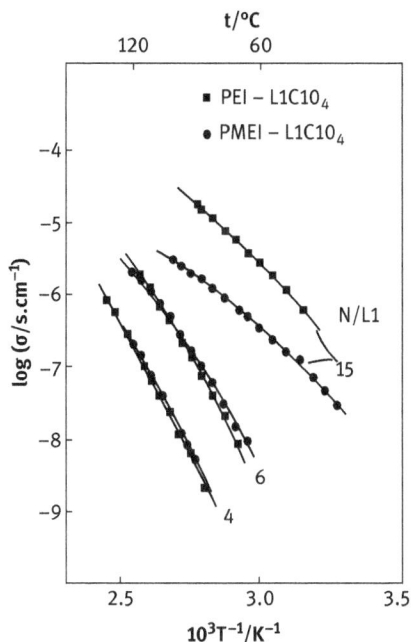

Fig. 5.35: Temperature dependence of ionic conductivity for PEI:LiClO$_4$ and PMEI:LiClO$_4$. Similar conductivities are observed for both hosts at high salt concentrations (N/Li = 4–6) and higher conductivity for PEI compared to PMEI at low salt concentrations (N/Li = 15). Adapted from [154], Copyright 1993, with permission from Elsevier.

hygroscopic, PPI is relatively anhydrous, as determined by IR, which suggests an exclusion of water during crystallization of PPI. In addition, IR also shows weaker hydrogen bonding in PPI compared to PEI [169]. Linear PPI can be synthesized by cationic ROP of 2-ethyl-5,6-dihydro-4-H-1,3-oxazine followed by a hydrolysis step to remove the pendant acyl group. At low LiCF$_3$SO$_3$ concentrations, the degree of crystallinity is reduced from 100% for pure PPI to 43% for N/Li = 20. At higher concentrations above N/Li = 5, an amorphous SPE is formed. Regarding ionic conductivity, similar values to those for PEI have been obtained, <10^{-7} S cm^{-1} at room temperature and >10^{-5} S cm^{-1} at 50 °C [169].

Coordination of alkali metal ions can also occur through dynamic labile metal–ligand bonds (Fig. 5.36a). Polymers containing tethered imidazole ligand moieties are able to form dynamic cross-links with cations, thereby decoupling the mechanical properties from ionic conductivity. The physical cross-links increase the mechanical properties of the matrix while the labile nature of the M–L bonds promotes the ion transport through the material [170, 171].

A polymer containing PEO as the polymer backbone with imidazole pendant units (PIGE, Fig. 5.33) has been synthesized through a combination of ring-opening anionic copolymerization followed with thiol–ene click chemistry. The resulting polymer is amorphous (T_g –33 °C) and the T_g increases upon addition of Ni(TFSI)$_2$, indicating that the metal–ligand complexes impose restrictions on polymer segmental motion [170]. It is also able to dissolve other metal-TFSI salts (M = Li$^+$, Cu^{2+}, Zn^{2+}, Fe^{3+}) and the ionic conductivity of PIGE:MTFSI is similar for all cations at imidazole/M = 10 (or 4.8 wt%

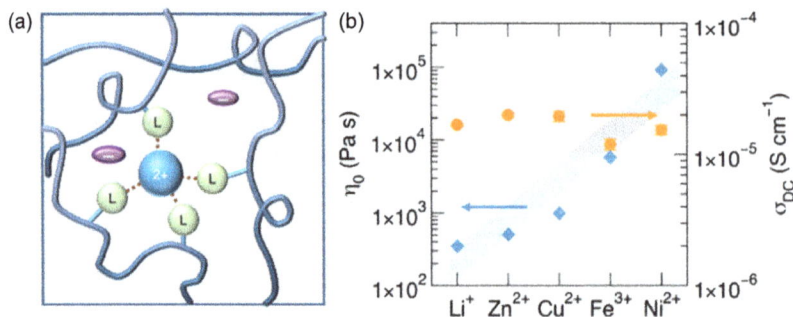

Fig. 5.36: (a) Schematic representation of a dynamic metal–ligand coordination. (b) Zero-frequency viscosity and ionic conductivity dependence on metal cation. Reprinted with permission from [171]. Copyright 2018 American Chemical Society.

for LiTFSI) being 10^{-5} S cm^{-1} at 70 °C. The transference numbers for the polymer with and without the imidazole groups are 0.15 and 0.19, respectively. The slightly higher t_+ of the latter is attributed to a weaker Li$^+$–polymer interaction in the presence of imidazole, as the Li$^+$ are no longer as strongly coordinated to PEO because the cations have preferential coordination to imidazole units, as shown with 2D NMR experiments. For PIGE:LiTFSI, the polymer zero-frequency viscosity (related to the dynamic cross-links) is similar to the pure polymer, indicating that each Li$^+$ interacts with only one imidazole (not forming temporary cross-links) or that the timescale is short compared to the polymer dynamics. For the other cations, however, the polymer zero-frequency viscosity changes several orders of magnitude with negligible changes in ionic conductivity, thus indicating a decoupling of both properties (Fig. 5.36b) [171].

Despite the structural analogy of polyamines to polyethers, polyamines have not gained as much attention and interest. While the hydrogen bonding is beneficial for anion coordination that increases T_+, it also promotes crystallization of the polymer. Even for amorphous polyamine-based SPEs, the ionic conductivity is still rather low. Further understanding of the polymer–salt interactions and ion transport mechanisms would be required to design new polymer hosts to be used in practical battery devices.

5.5 Polyalcohols

The pendant hydroxyl group of polyalcohols clearly is sufficiently electron-rich to have complexing capabilities for metal cations, and this type of polymers is thereby theoretically a useful class of host materials for Li- and Na-salts, leading to the formation of SPEs. Furthermore, many polyalcohols are produced on very large scales, and the resulting materials could have clear advantages in terms of cost. While low-molecular-weight alcohols display limited electrochemical stability, their macromolecular analogues should be somewhat more stable due to the more uniform local electron density, and the reaction products, which could form on the electrode surfaces, might have limited solubility back into the SPE matrix. The hydroxyl groups may also interact not only with metal cations, but also with their negatively charged counterions. Moreover, their hydrogen-bonding capabilities give them interesting properties, which can be useful for mechanical properties and different "self-healing" capabilities.

So far, a rather limited number of polyalcohols have been explored for SPEs (see Fig. 5.37 for examples), where PVA clearly dominates the existing literature on this subject. Some interesting properties have been seen, not least their ability to dissolve substantial amounts of salt and that the ionic conduction to a large degree seems to be decoupled from the polymer backbone mobility. In this sense, the conductivity often displays a straight line in the Arrhenius diagram (log σ vs $1/T$), and the conductivity can be substantial also below T_g. Furthermore, in contrast to for example PEO, the conductivity also increases with salt concentration over the entire solubility range, and there does not seem to exist any conductivity decrease due to limitations in mobility [172].

The typical procedure for synthesizing PVA is starting from vinyl acetate, which can be transformed to poly(vinyl acetate) (PVAc) through free-radical polymerization [173]. The acetate groups are thereafter hydrolyzed to attain PVA, but often not fully, which results in a fixed ratio between hydroxyl and acetate groups known as the *degree of hydrolysis* (or alternatively, as the degree of acetylation) [174]. This approach also normally renders a wide distribution in molecular weight (a polydispersity index, PDI, >2) and an atactic configuration of the pendant functional groups. The absence of stereoregularity does not, however, result in a largely amorphous polymer as one might expect. Instead, the large number of possible hydrogen bonds renders a well-ordered crystalline structure in a monoclinic unit cell [175]. When salt is dissolved into the PVA matrix, on the other hand, the resulting films become amorphous and transparent [172]. Similarly, dissolution of a salt into PVA also renders a substantial lowering of the T_g [176]. The hydrogen bonding otherwise constitutes the background to a relatively high T_g of PVA at 85 °C. A similar effect has also been observed for hydroxyl-functionalized SPEs based on polycarbonates, which was attributed to the ability of the polymer to interact strongly with the anion, and not only with the cation [109]. As such, high transference numbers could be expected from polyalcohols.

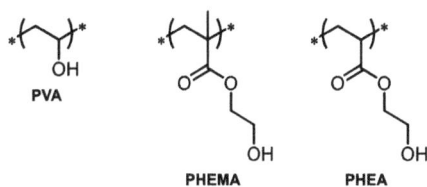

Fig. 5.37: Polyalcohols used as SPE host materials: poly(vinyl alcohol) (PVA), poly(hydroxyethyl methacrylate) (PHEMA) and poly(hydroxyethyl acrylate) (PHEA).

One problematic feature of PVA is that the polymer is not soluble in any typical volatile organic solvent that is easily evaporated after casting. Instead, the high-boiling solvent DMSO is typically used. It has shown to be notoriously difficult to fully eliminate all solvent after casting, most likely due to strong interactions between DMSO and PVA itself, or DMSO and the salt. In studies where the trace amounts of DMSO after casting have in fact been quantified, residues of DMSO in the range of 3.5–12% have been detected [172, 176, 177]. As discussed in Chapters 1–4, these solvent residues contribute to conductivities, which are clearly beyond what should otherwise be possible, and similarly to some polynitrile examples – and non-coordinating polymers for that matter – constitute classic examples of how such conductivity data can easily be misinterpreted. To overcome the problems associated with solvent residues, solvent-free hot-pressing techniques can be applied to produce SPEs. This was successfully applied for PVA-based SPEs as PVA:LiCF$_3$SO$_3$ (see Fig. 5.38) [176] and more recently for LiTFSI-based analogues [177]. In both these systems, an Arrhenius-type behavior was seen for the temperature dependence of the ionic conductivity – but orders of magnitude lower than for the corresponding solvent-cast samples. This clearly shows the key functionality of solvent residues for useful electrolyte functionality.

Still, the ion transport mechanism in polyalcohols has not been explored to a very high degree and uncertainties exists. As also seen in Fig. 5.38, decoupled-type conductivities below T_g can be observed, which is not easily explained only by the remaining solvent fraction. Also the polymers PHEMA and PHEA, which possess hydroxyl groups positioned quite distant from the main chain as compared to PVA, display similar behavior. One approach to achieve further insight into the transport phenomenon has been using ^{7}Li NMR, which was employed for the PVA:LiCF$_3$SO$_3$ system, using PVA with different degrees of hydrolysis [178]. As the lithium ion mobility increases, the corresponding NMR peaks become sharper. By investigating the NMR signals below T_g, such effects could clearly be seen. Moreover, sharper NMR signals and lower activation energies could be observed for lower degrees of hydrolysis of PVA, which is consistent with the measured higher conductivity data. It is possible for these PVA-based systems that the remaining acetyl groups induce a breakdown of the symmetry, and prevent stable formation of a hydrogen-bonding network, which can explain this effect. Consequently, the cation transport in the PVA:LiCF$_3$SO$_3$ system has been suggested to occur by a hopping mechanism, or alternatively through a secondary polymer relaxation process where also proton conduction plays a role. On the other hand, since operational Li-metal batteries have been constructed using PVA-based electrolytes [177], it seems unlikely that there exists any dominating contribution from proton conduction – if so, these hydrogens would be rapidly consumed by the lithium metal during cycling.

There has only been limited application of PVA-based electrolytes in batteries. It was shown that more novel salts such as LiTFSI display an increased conductivity as compared with PVA:LiCF$_3$SO$_3$, but again that DMSO (10 wt%) residues are necessary for satisfactory conductivity. Nevertheless, a functional Li-metal | PVA:LiTFSI (DMSO) | LFP battery operating at 60 °C was constructed, which displayed a stable capacity (136 mAh g^{-1}) but for low cycling rates and a rather limited number of cycles [177].

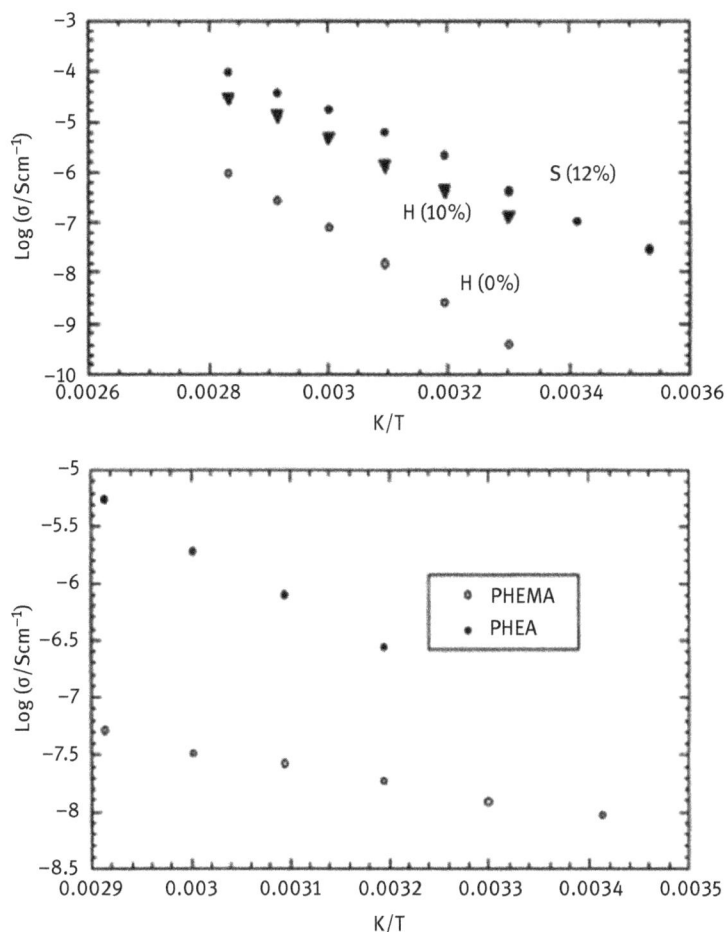

Fig. 5.38: Ionic conductivity of SPEs based on different hydroxyl-functional polymers (top: PVA; bottom: PHEMA and PHEA) with LiCF$_3$SO$_3$ salt. The amount of DMSO residues after hot-pressing (H) or solvent-casting (S) are given in parentheses in the top figure. Adapted from [176], Copyright 1998, with permission from Elsevier.

Moreover, a system of DMF-cast PVA:LiBOB electrolytes with up to 50 wt% salt was explored in a lithium–oxygen battery, and a respectable conductivity maximum of 2.85×10^{-4} S cm^{-1} at 40 wt% salt was reported as a result of a significant reduction in both T_g and crystallinity when salt was added [179].

It could be speculated, however, that the much-reduced conductivity for polyalcohols when fabricated fully solvent-free constitutes the last nail in the coffin for utilizing these materials as conventional solvent-free SPE hosts. On the other hand, as a polymer base for different gels, polyalcohols could well be excellent host materials. Moreover, there could also be an increasing use of PVA as an electrode binder, where its ion-coordinating capabilities can render useful functionalities for battery applications [180, 181].

5.6 Polymerized ionic liquids and ionomer concepts

Besides the conventional types of SPEs based on salt mixed with polymers, which have been described in the preceding chapters and sections, there are other types of SPEs that contain ions incorporated directly into their structures. Thereby, they are members of the family of *polyelectrolytes*. Polyelectrolytes are polymers with dissociating groups in their repeating unit, thus, the polymer backbone is charged and it has counterions ionically bonded to them to compensate these backbone charges. As these polymers contain free ions, they are intrinsically ion conductors, which make them interesting candidates to be used as SPEs in batteries. The broader area of polyelectrolytes has a rich literature, but primarily treats the conventional polyelectrolytes that are soluble in aqueous solutions with the ions highly dissociated. The use of liquids, and especially electrochemically reactive H_2O, renders those materials out of the scope for this book. Moreover, depending on the chemical structure of the backbone and the number of ionic centers, there exist several different types of polyelectrolytes. This chapter will focus on those relevant for battery applications, that is, *ionomers* and *polymerized ionic liquids*.

Ionomers constitute a subclass of polyelectrolytes comprising polyelectrolytes that combine electrically neutral and ionized groups in the polymer backbone distributed randomly or regularly. Ionomers are considered to have less than 10–15% ionic groups, and most ionomers are insoluble in water [182, 183]. The first ionomer was produced by DuPont in the early 1960s; a random copolymer consisting of poly (ethylene-*co*-methacrylic acid), called Surlyn®. Since then, many other structures and applications have been found for ionomers, including solid-state batteries.

Another subclass of polyelectrolytes is polymerized ionic liquids or poly(ionic liquid)s (PILs) whose repeating unit is an ionic liquid species. An ionic liquid is basically a salt which melts below 100 °C. Normally, they have high ionic conductivity, good thermal stability and nonflammability properties. PILs combine the properties of polyelectrolytes and ionic liquids, primarily being solid and a reasonable ionic conductor. In contrast to conventional polyelectrolytes, most PILs are soluble and dissociate in polar organic solvents [184, 185]. The concept of PILs was originally proposed in the 1990s to introduce a new class of solid electrolytes that could potentially substitute ionic liquids in electrochemical devices. Since then, the application of PILs as polymer electrolytes in energy storage has gained a lot of interest thanks to their versatility, solubility of the salt, high thermal stability and inherent ionic conductivity when the counterion is a cation [186–189].

Both ionomers and PILs contain ionic centers in their structures (Fig. 5.39); polycations if the backbone is positively charged with associated counteranions, and polyanions if negatively charged with countercations. For most battery applications, polyanions are more interesting because they contain the desired mobile metal cation, which makes

Conventional SPEs

Polyelectrolytes

Poly(ionic liquid)s

Ionomers

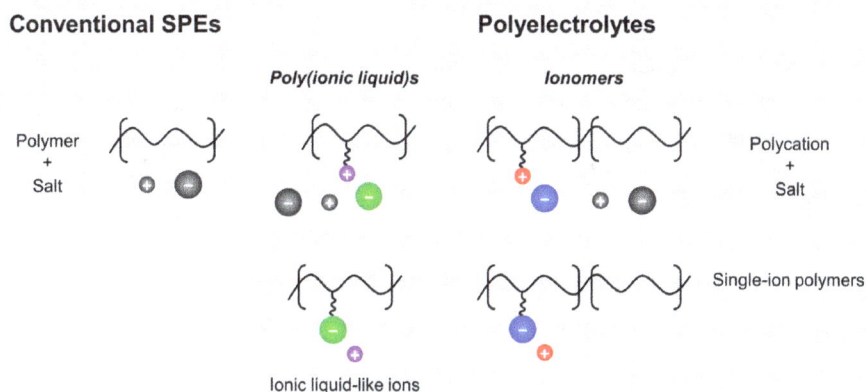

Fig. 5.39: Graphical representation of conventional SPEs and polyelectrolytes focusing on the specific types of PILs and ionomers.

addition of any extra salt unnecessary. A polyanion with, for example, a Li^+ is known as a "single-ion polymer electrolyte" because the electrochemically useful cation is the only long-range mobile ion in the system. Their transport number t_+ is thereby very close to 1, which prevents the formation of concentration gradients during battery operation. In contrast, polycations require the addition of a salt, similar to other types of SPEs, to be used as electrolytes for battery applications. While the most common ionomers are polyanions – because more applications have been found for this type of polymers – the most common PILs are polycations due to their easier synthesis.

5.6.1 Chemical structure of PILs and ionomers

Both PILs and ionomers contain charged groups and mobile ions in their structure. However, there are many other features that differentiate them, the main ones being the nature of the ions and the amount of charged species in the polymer chain. In both cases, there is an endless number of polymer backbones and ionic species that can be combined, and these will determine the final properties of the SPE.

There are three main parts in the structure of PILs that should be considered: the polymer backbone, the ionic species and the spacer connecting both of them. The most common type of PIL is a polycation with anionic pendant groups, primarily due to its easier synthesis route. Some examples of cations in the polymer backbone are imidazolium, pyrrolidinium and ammonium (Fig. 5.40a), the counteranions generally being TFSI, FSI, dicyanamide, BF_4^- or PF_6^- (Fig. 5.40b). Similar to other SPEs, the most commonly used salts with PILs are based on TFSI anions. Anionic PILs have been less explored than the cationic analogues because of their more difficult synthesis. The useful cationic counterions are then, for example, Li⁺ for lithium bat-teries and Na⁺ for sodium batteries. Typical anionic groups on the backbone are TFSI, carboxylate, phosphonate or sulfonate (Fig. 5.40c). The polymerizable units of the polymer backbone are usually vinyl, styrene, (meth)acrylate, (meth)acrylamide, vinyl ether and norbornene. In addition, the spacer between the polymerizable unit and the anion or cation of the polymer backbone could be short or long chains of ethylene or ethylene oxide fragments.

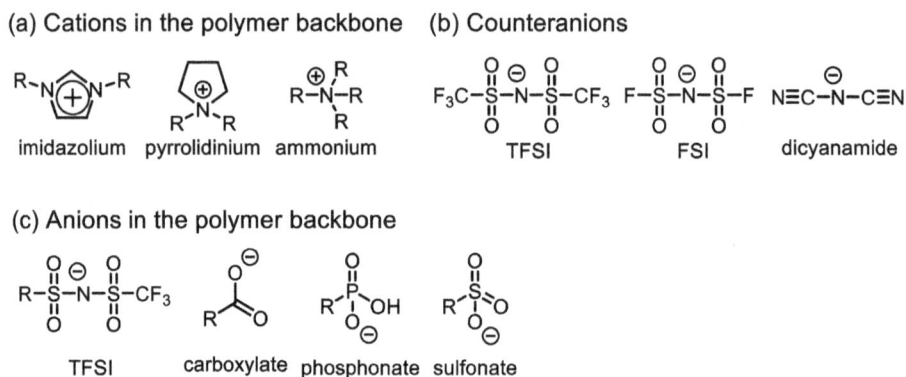

(a) Cations in the polymer backbone   (b) Counteranions

imidazolium   pyrrolidinium   ammonium   TFSI   FSI   dicyanamide

(c) Anions in the polymer backbone

TFSI   carboxylate phosphonate sulfonate

Fig. 5.40: Chemical structures of (a) cations in the polymer backbone, (b) counteranions and (c) anions in the polymer backbone of PILs and ionomers.

Ionomers combine charged and electrically neutral units in their structures. Histori-cally, polymers with 10–15% charged units were termed ionomers [190–192]; how-ever, the same term is being used also for higher amounts. Important features of ionomers are the chemical structure of the polymer backbone, the nature of the counterion and the concentration and distribution of each part. These will deter-mine the possible synthesis routes as well as the final properties of the ionomers. In general, in this category of polyelectrolytes, polyanions are more common and con-sidered to be more important from a practical standpoint than polycations, and this is also true for battery applications. Among the different types of charged species in ionomers, those containing carboxylic acid and sulfonate groups have been the most widely explored since their introduction in the 1960s [193]. More recently, an-other type of anion based on TFSI has been used [194], because the high delocaliza-tion of its negative charge favors dissociation of small cations, thereby enhancing the ionic conductivity [195]. The ion–ion interactions are otherwise strong in ionomers, which lead to strong complexation that hinders ion transport. Therefore, the nature of the electrically neutral units is important as they will contribute to the dissolution of the countercations – if they include Lewis basic groups – thus increasing the ionic conductivity. The most common motif is PEO, but other examples are polystyrene, polyurethane, polyethylene and polyacrylate.

The most studied ionomers for solid polymer electrolytes are block copolymers containing a hard block and a soft block. The anionic groups could be placed either in the hard segment or in the soft segment. Most of the reported ionomers for polymer electrolytes have the ionic groups in the hard segment, typically polystyrene [196].

Polyanions from the PIL and ionomer families can be combined in the term "single-ion polymer electrolytes," most often used in the battery application perspective. They resemble anionic PILs because they usually contain similar anions based on TFSI and they are similar to ionomers because the polymer backbone combines anions and electrically neutral units. It is therefore not clear exactly where the line should be drawn between these types of polymers and ionomers. Thus, all polyanions discussed in this chapter will be referred to as single-ion polymer electrolytes irrespective of whether they belong to the PIL or ionomer family.

5.6.2 Synthesis of PILs, ionomers and single-ion polymer electrolytes

Regardless of if the target is a PIL, an ionomer or a single-ion polymer, there are two general synthesis routes that can be used (Fig. 5.41): (1) direct polymerization of monomers, and (2) chemical modification of existing polymers. The first method allows more freedom to choose a specific polymer structure, as well as straightforward synthesis of homopolymers and copolymers. However, it also involves a more complex monomer preparation route with different organic synthesis and purification steps. The second method is synthetically less complicated as it involves fewer synthetic steps, but is on the other hand limited in usefulness if more advanced structures are required.

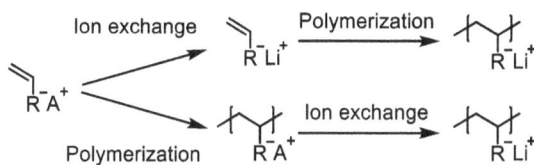

Fig. 5.41: General synthetic routes of charged polymers.

Both strategies involve a polymerization step that can be carried out through conventional and controlled radical polymerization, ring-opening metathesis polymerization, step-growth polymerization, etc. Another step is an ion exchange reaction usually from halide anions to another anion or from potassium or quaternary ammonium groups to lithium or sodium cations. In some cases, the chemical structure of the final polymer synthesized from direct polymerization or modified polymers could be the same and the choice of one approach or another will be based on its feasibility, as some polymers can only be synthesized through one of the routes.

Particularly for PILs, the monomer's structure should be similar to an ionic liquid but with one or more polymerizable units. For general ionomers, the monomers do not have any such requirement. Generally, the preferred synthesis route for both types of polymers is free-radical polymerization as it has high tolerance toward impurities, moisture and the presence of other active and functional groups. Examples of polymers synthesized by free-radical polymerization are LiPSsTFSI with a more delocalized anion compared to TFSI [197], Na-PSTFSI [198], copolymers of 2-acrylamido-2-methyl-1-propane-sulfonate (AMPS) and the sodium salt of vinyl sulfonate (NaVS) [199] (Fig. 5.42a). Other types of PSTFSI-based polymers have been synthesized with controlled radical polymerization methods such as RAFT [196], NMP [194] and ATRP [200]. Polycarboxylates [201], polyurethanes [202] and polyesters [203] have been prepared with step-growth polymerization (Fig. 5.42b).

(a) Polymers synthesized by radical polymerization

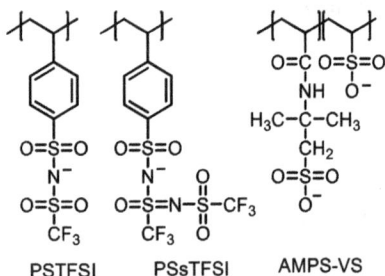

PSTFSI   PSsTFSI   AMPS-VS

(b) Polymers synthesized by step-growth polymerization

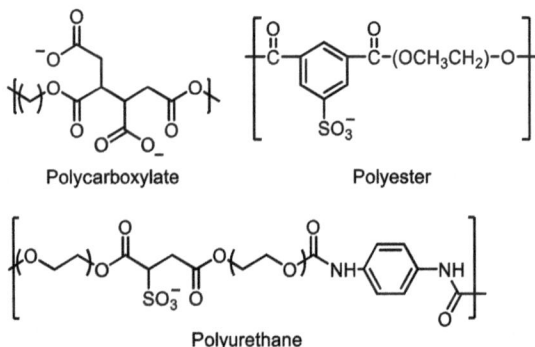

Polycarboxylate   Polyester

Polyurethane

Fig. 5.42: Chemical structures of polymers synthesized by (a) radical polymerization and (b) step-growth polymerization.

5.6.3 Properties of PILs, ionomers and single-ion polymer electrolytes

Similar to other polymer electrolytes and as a general rule, polymers with lower T_g have higher bulk conductivity, at least if strongly correlated to segmental mobility. The T_g of PILs and ionomers not only depends on the chemical composition of the polymer backbone but also on the type of counterions. For example, a neutral poly (2-(dimethylamino)ethyl methacrylate) exhibits a T_g at 19 °C, and after quaternization and introduction of anions the T_g increases due to the introduction of ionic aggregation. A larger TFSI with weaker interactions with the polycation featured a lower T_g at 38 °C compared to the T_g of PF_6^--based PIL at 164 °C [204]. Copolymers of PEG with dimethyl 5-sulfoisophthalate salt show slightly higher T_g values when changing from Li^+ to Na^+ and to Cs^+, indicating that the larger size of the cation hinders the chain mobility [205]. Comparing the T_g of the Na^+ ionomer to PEO with NaTFSI and NaClO$_4$ SPEs, at low Na^+ content the T_g values are similar for the three cases; however, a much larger T_g is seen for the ionomers (20 °C) than for PEO with NaTFSI (−30 °C) and PEO with NaClO$_4$ (−20 °C) at 0.1 $[Na^+]/([Na^+]+[EO])$ due to the presence of ionic domains in the ionomers [203].

Generally, PILs and ionomers are noncrystalline amorphous polymers probably because the presence of mobile counterions hinders the crystallization process [185]. As an example, the neutral analogue of the aforementioned PEG-dimethyl isophthalate polymer is highly crystalline and a T_g cannot be detected, while the ionomer with the pendant ions shows a T_g around 20 °C [203].

The ionic conductivity of PILs and ionomers is usually rather low compared to other SPEs. For example, a vinylimidazolium-TFSI PIL featured an ionic conductivity of 10^{-7} S cm^{-1} at room temperature; however, the mobile species is the anion, which is not relevant for battery applications. When adding LiTFSI to the system, the ionic conductivity increases to 10^{-6} S cm^{-1} at room temperature. Free volume is gained by adding the salt which facilitates the ion mobility and thereby increases the ionic conductivity [206]. Similarly, a polyanion ionomer had two orders of magnitude lower ionic conductivity than the analogous neutral polymer with addition of LiTFSI [207]. Nevertheless, it is important to mention that in the salt-doped polymer both ions are mobile and contribute to the ionic conductivity, whereas in the ionomer only the cation is mobile, and thereby a lower ionic conductivity would be expected. Furthermore, in the previous example clusters of ions detected by SAXS were formed only in the ionomers, which do not contribute to the conductivity either. Such clustering was not detected in the salt-doped polymers [207].

Similar to other SPEs, multiple approaches have been developed to improve the ionic conductivity and electrochemical properties of PILs and ionomers. One approach to increase the ionic conductivity without affecting – or perhaps even improving – the mechanical stability is to develop block copolymers incorporating a soft (ion-conducting) block together with a hard (mechanically robust) block. This strategy is relevant also for conventional SPEs but, in the case of PILs, ionomers and single-ion polymer electrolytes, the position of the ionic groups also needs to be considered. This behavior has for example been studied for polyurethane–polyethylene oxide copolymers [208, 209]. In polyurethanes, the hydrogen bonds can create a physical cross-link between urethane linkages and the copolymer microphase separated into hard and soft blocks. This has shown that placing the carboxylic ionic groups in the hard polyurethane blocks (Fig. 5.43a) increases the ionic conductivity but reduces the mechanical stability, because the ions in the hard segment compete for the hydrogen bonds of the urethane unit, thereby preventing discrete microphase separation. Placing the anions in the soft segment (Fig. 5.43b), however, does not lead to phase separation either. If instead chain extenders are incorporated to increase the hard segment required for microphase separation, it is possible to obtain phase-separated materials with high storage modulus and high ionic conductivity at elevated temperature (150 °C). Since the ionic conductivity at room temperature is too low to be used in batteries [202], further development of these systems is required to obtain both high mechanical stability and ionic conductivity at room temperature.

(a) Ionic groups in the hard segment

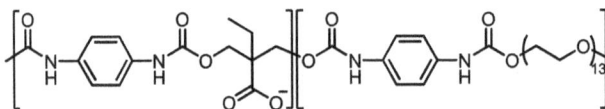

(b) Ionic groups in the soft segment

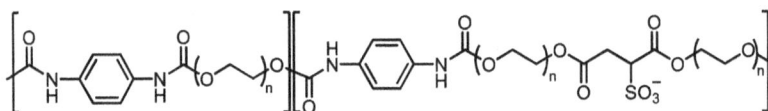

Fig. 5.43: Chemical structures of ionomers containing ionic groups in (a) the hard segment and (b) the soft segment.

Another common hard block is polystyrene, which has been modified with a TFSI-analogous anion to build a single-ion block copolymer with PEO as soft block. Michel Armand and coworkers were among the pioneers to develop this type of polymers obtaining fairly high ionic conductivity (1.3×10^{-5} S cm^{-1} at 60 °C and around 3×10^{-5} S cm^{-1} at 90 °C), high transport number (>0.85) and high mechanical stability (a tensile strength of 10 MPa at 40 °C) [194]. When the morphology–conductivity relationship of this type of SPE was later studied, it was reported that the material below 50 °C presents a microphase-separated structure with crystalline PEO-rich domains and glassy PS-TFSI-rich domains, where ionic clusters are located. Above 50 °C, when the morphology is disordered and PEO and PS-TFSI are intimately mixed, the ions are no longer in clusters and the ionic conductivity is higher (3.8×10^{-4} S cm^{-1} at 90 °C) [210]. The PEO block can also be incorporated as side chains [196]. However, the ionic conductivity cannot be straightforwardly compared because the molecular weight of the final polymers and the ratio between blocks is different. In fact, additional polystyrene blocks were included to promote phase separation forming a triblock copolymer A-B-A; A being polystyrene and B a copolymer of styrene with TFSI and grafted PEO. Increasing the concentration of the central block (PS-TFSI and PEO) results in an increase in ionic conductivity as the number of charge carriers increased, as well as an increase in morphological phase separation and channel connectivity. The ionic conductivity was 10^{-5} S cm^{-1} and the elastic modulus was in the range of 10–100 MPa at 90 °C [196]. In addition, MD simulations of these system showed that the STFSI side-chain has very little effect in the Li-ion diffusion mechanism. In fact, increasing the length of the PS-TFSI block leads to a less flexible overall polymer chain, with limited segmental motion thereby limiting the Li mobility [211]. This type of polymers has also been reported with sodium cations, and a copolymer of PS-TFSI with a polyacrylate showed higher ionic conductivity than the PS-TFSI homopolymer, because the acrylate units facilitate the cation dissociation [198].

Besides styrene, TFSI-based anions can be incorporated into a methacrylate-based monomer, the latter having slightly higher ionic conductivity (6.1×10^{-5} S m^{-1} at 90 °C) compared to the styrene analogue (3.3×10^{-5} S cm^{-1} at 90 °C) (Fig. 5.44) [212, 213]. Incorporating the PEO units as side chains with a PEG-methacrylate results in higher ionic conductivity. As expected for a single-ion polymer electrolyte, a high transference number of 0.91 is obtained for this polymer [214].

The effect of the polymer nanostructure has also been studied for polycation-based PILs, where the interchain distance determined by the length of the alkyl chain in the cation influences the ionic conductivity. Similar to the PISE materials described in Section 2.3, the intermolecular anion hopping as ion conduction mechanism is correlated to the distance between chains. Longer alkyl chains hinder the anion hopping from one chain to another, thereby decreasing the ionic conductivity [215]. Incorporating LiTFSI in this type of system slightly increases the T_g due to the formation of ionic interactions between the polymer chains. However, the ionic conductivity is generally increasing due to the gain in free volume that facilitates the ionic mobility [206]. This shows a way to control the polymer's nanostructure and to increase the ionic conductivity in cation-based PILs.

Usually, the electrically neutral groups in this type of polymers have relatively low dielectric constant, similar to the majority of polymer electrolyte host materials (see Chapter 2). In absence of solvent, the ions from the other charged blocks therefore tend to aggregate locally, holding together different chains and forming temporarily cross-linked units, which contribute to improving the mechanical properties of the polymer. Increasing the concentration of the ionic groups leads to the formation of percolated ionic aggregates where ion transport can take place. Thus, forming these percolated ionic aggregates is another way to decouple from segmental motions without compromising the mechanical properties. It has also been suggested from molecular dynamics simulations that the transport of lithium ions occurs within these percolated

PSTFSI-b-PEO-b-PSTFSI

PMATFSI-b-PEO-b-PMATFSI

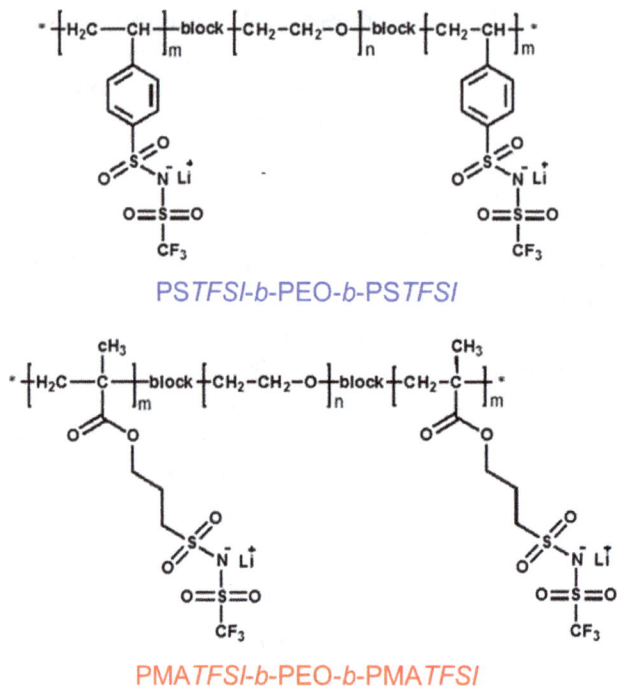

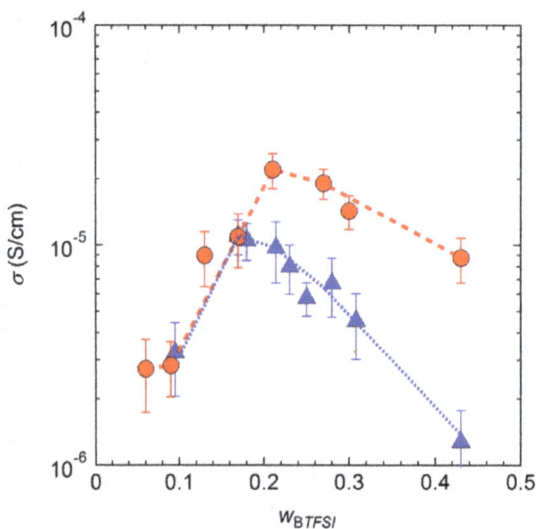

Fig. 5.44: Isothermal ionic conductivity at 60 °C as a function of the weight fraction of BTFSI block, w_{BTFSI}. The symbols correspond to (triangles) PSTFSI- and (circles) PMATFSI-based electrolytes. The dashed lines are guidelines depending on the nature of the BTFSI block. Adapted from [212], Copyright 2018, with permission from Elsevier.

ionic aggregates [216, 217]. However, in the absence of percolation (for partially neutral-ized ionomers or lower concentrations of lithium cations), ion transport occurs through the rearrangement of lithium ions inside an aggregate and through merging and break-ing up of ionic aggregates. Therefore, decreasing the spacing between ions and increasing the ion density allows for larger, percolated ionic aggregates [216, 217]. These findings have also been confirmed experimentally, showing that the highest neutralization level, that is, the highest number of lithium cations, and shorter alkyl spacers between ionic groups, leads to higher conductivity (Fig. 5.45). This is because at high neutralization, the ion transport is decoupled from the polymer backbone dy-namics. However, the strong electrostatic interaction between the carboxylate anions and the lithium cations is still limiting the ionic conductivity (below 10^{-9} S cm^{-1} at 40 °C) [201].

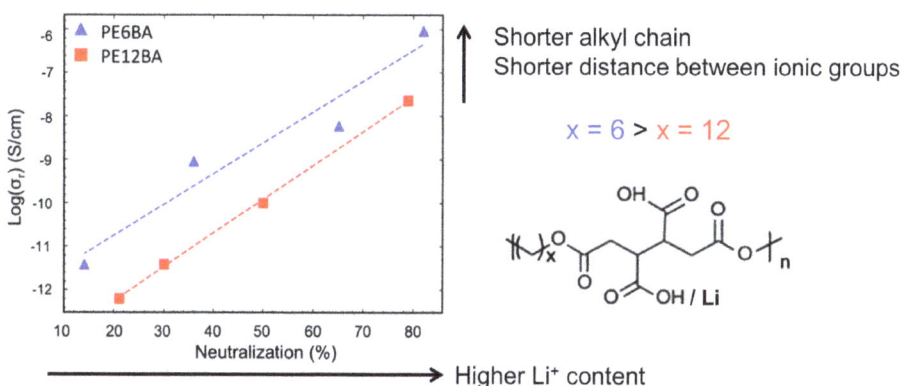

Fig. 5.45: Effect of the Li$^+$ content and distance between ionic groups on the ionic conductivity of carboxylate-based ionomers at T_g + 20 K. Adapted with permission from [201]. Copyright 2020 American Chemical Society.

Another approach to increase ionic conductivity is through the incorporation of bulky organic co-cations such as alkyl ammonium. These co-cations reduce the strong ion aggregation, act as plasticizers and increase the free volume by lowering the T_g and then increasing the ionic conductivity [218, 219]. However, the ionic conductivity val-ues are still too low for practical application in batteries, ranging between 10^{-10} S cm^{-1} at 30 °C and 10^{-5}–10^{-6} S cm^{-1} at 100 °C [199, 220].

There are many different anions that can be incorporated into polymers. For ion-omers, most of the reported examples contain carboxylate or sulfonate groups, proba-bly because those have been common for proton-conducting membranes in fuel cells. However, lithium or sodium ions are too tightly associated with those anions and therefore the mobility and hence conductivity are limited [217]. Other type of reported anions are (trifluoromethyl)sulfonyl acrylamide [221], and boron-based anions [222], but these also render low ionic conductivity. As shown previously, a better choice of anion is TFSI-based as the negative charge is more delocalized and the ion pairs are more dissociated. Therefore, in this line of thought, developing new delocalized anions that further enhance the dissociation of the ions could be an important strategy to improve the ionic conductivity of ionomeric SPEs (Fig. 5.46). In this context, one of the S=O bonds from TFSI has in a recent study been replaced with another $=NSO_2CF_3$ group, thereby creating a "super-delocalized" anion (Fig. 5.46a) with higher ionic conductivity compared to its analogous PS-TFSI-based polymer (Fig. 5.46b) [197]. Another example of a highly delocalized anion that is more sustainable (as it is fluorine-free) is a dicyanomethide-based anion (Fig. 5.46c). This has been proposed due to the easier dissociation of Li cations from the strong electron-withdrawing but poorly chelating cyano groups. This anion has been incorporated into a polystyrene-based backbone and blended with PEO. This strategy does not lead to the crystallinity of PEO being completely removed, and the polymer electrolyte features an ionic conductivity of 10^{-7} S cm^{-1} and a lithium transference number of 0.95 at 70 °C (Fig. 5.46d) [223].

The electrochemical stability of SPEs will depend on both the salt and the polymer used. Single-ion polymers, however, do not contain an additional source of mobile anions prone to be degraded on the surface of the electrodes. Therefore, this type of SPE could render a higher electrochemical stability window compared to conventional SPEs as the anions cannot migrate to the reactive surface [194]. For example, a polystyrene-TFSI/PEO block copolymer features an enlarged electrochemical stability window (up to 5 V vs Li$^+$/Li) [194] compared to PEO (up to 3.8 V vs Li$^+$/Li) [224]. Copolymerizing a methacrylate-based TFSI with PEG-methacrylate results in a slightly lower electrochemical stability window up to 4.2 V versus Li$^+$/Li [214]. Similar results have been obtained when comparing block copolymers of PEO with PS-TFSI or with polymethacrylate-TFSI. The latter showed a reduced electrochemical stability window of 4.0 V versus Li$^+$/Li compared to 4.5 V versus Li$^+$/Li for the former. This has been ascribed to the better compatibility between PEO and polymethacrylate-TFSI that leads to an electrochemical stability window closer to that of the low stability values associated with the PEO homopolymer [212, 213]. However, it is important to mention in this context that there are several factors that prevent accurate determination of the electrochemical stability window (see Chapter 3).

5.6.4 Application in batteries

In the scientific literature, the focus of PILs and ionomer materials has primarily been on the synthesis of new structures and understanding their ion transport properties and mechanisms. The fundamental understanding of their behavior as SPEs is required to optimize their structures and for their proper implementation in battery devices. In fact, the overall electrochemical performance of these materials is rarely investigated. This indicates that besides all the advantages of these systems, more efforts have to be made in order to further improve their properties to reach an

(a)

(b)

(c)

(d)

LiPSsTFSI

LiPSDM

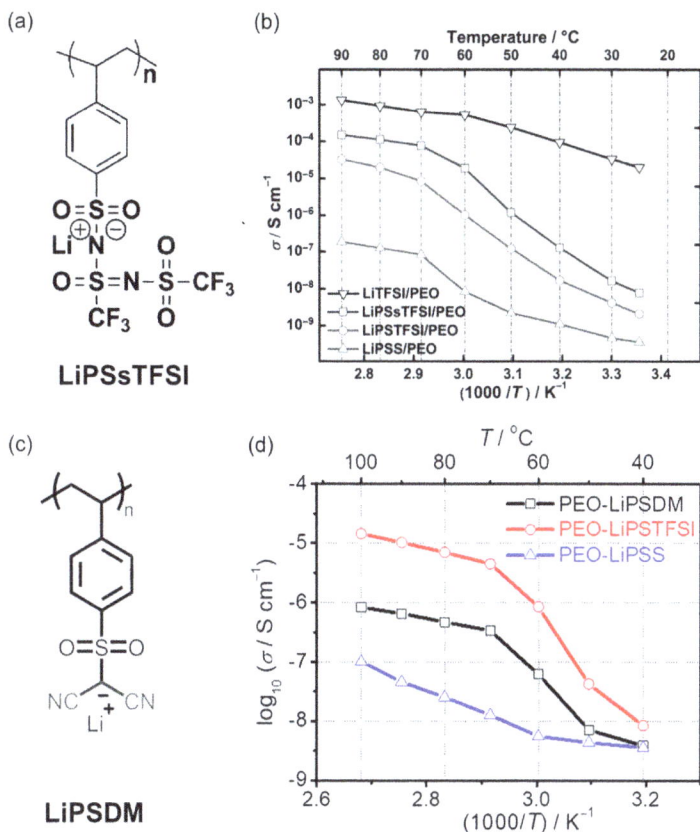

Fig. 5.46: Chemical structure of new delocalized polyanions (a) LiPSsTFSI and (c) LiPSDM. The temperature dependence of ionic conductivities for (b) PEO-LiTFSI, PEO-LiPSsTFSI, PEO-LiPSTFSI and PEO-LiPSS, reprinted with permission from [197] and (d) PEO-LiPSDM, PEO-LiPSTFSI and PEO-LiPSS (lithium poly(4-styrenesulfonate)). Reprinted with permission from [223]. © 2016 & 2020 Wiley-VCH.

application in real battery systems. However, a few pioneering examples of using these types of solid-state polymer electrolytes in true battery devices can be noted.

A polystyrene-TFSI/PEO block copolymer was implemented in a lithium metal battery as solid polymer electrolyte as well as a binder in an LFP cathode. This battery obtained good electrochemical performance between 60 and 80 °C (160 mAh g^{-1} discharge capacity at C/15) and a well-defined potential plateau up to a rate of C/2 with stability for up to 80 cycles (Fig. 5.47) [194]. A similar single-ion polymer but composed of polymethacrylate-TFSI instead of using a styrene monomer showed higher ionic conductivity. When implemented in lithium batteries with LFP as cathode, the performance was also better. A discharge capacity of 153 mAh g^{-1} was obtained at C/10 at 70 °C and stable cycling was obtained at C/2 for 300 cycles at 70 °C (Fig. 5.48)

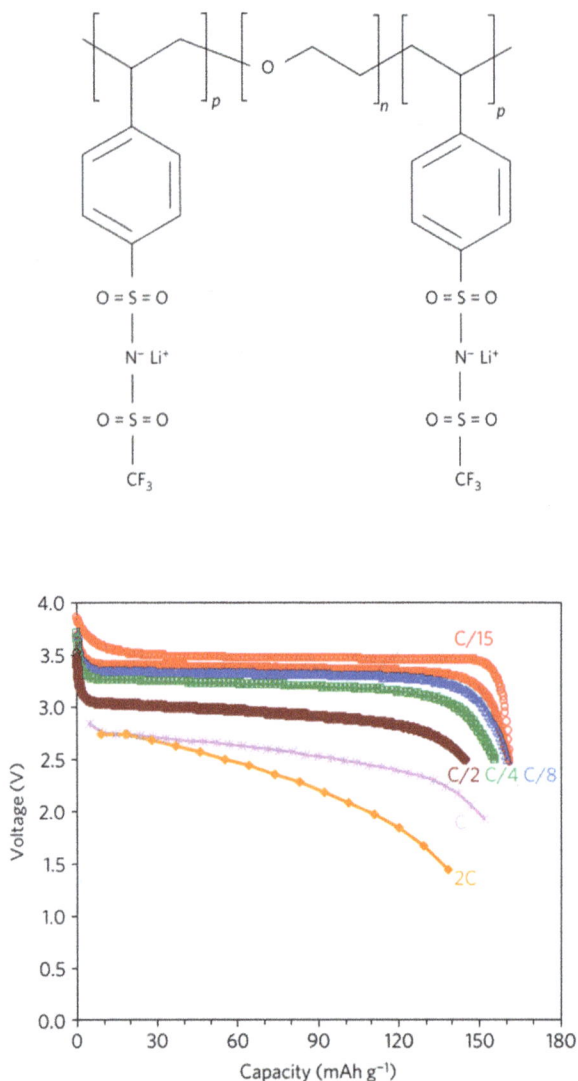

Fig. 5.47: Chemical structure and electrochemical performance of Li | PSTFSI-PEO-PSTFSI | LFP at 80 °C at different rates from C/15 to 2 C. Adapted by permission from Springer Nature Customer Service Centre GmbH [194], 2013.

[213]. Incorporating PEO as side chain led to slightly lower ionic conductivity and also less useful electrochemical performance in batteries with an LFP cathode. Although the initial discharge capacity was high (130 mAh g^{-1}) at 70 °C and C/15, the battery suffered a continuous capacity fade reaching 60 mAh g^{-1} after 90 cycles [214, 225].

Besides block copolymers, a blend of PSTFSI and PEO has been mixed together with 2 wt% LiFSI as additive to enhance the interaction between both polymers,

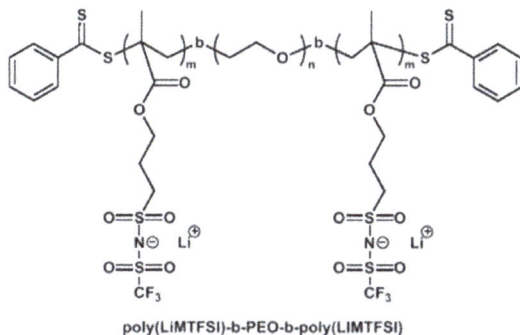

poly(LiMTFSI)-b-PEO-b-poly(LiMTFSI)

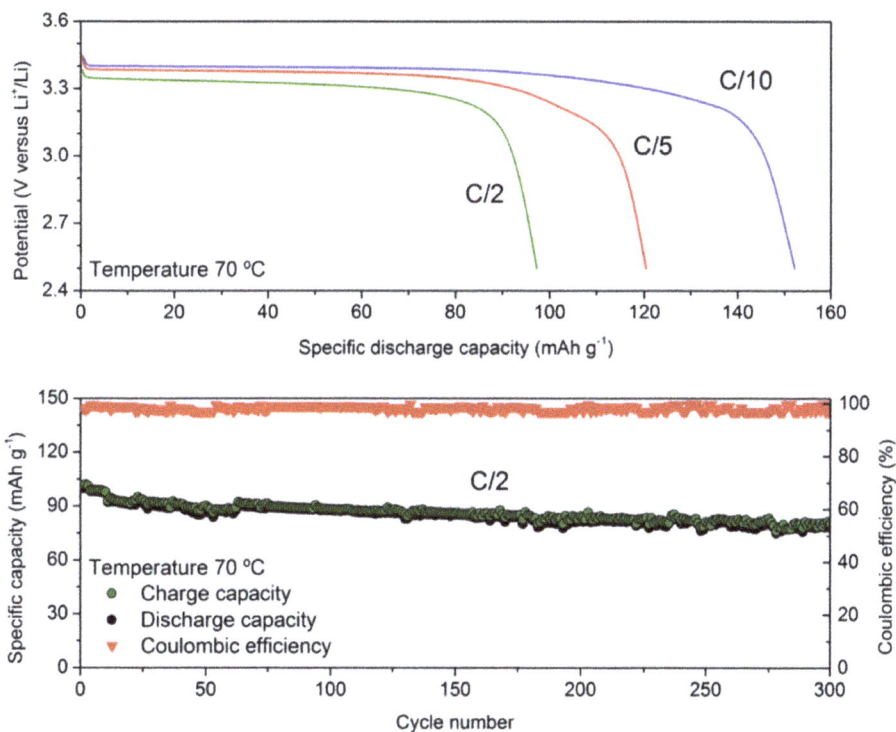

Fig. 5.48: Chemical structure and electrochemical performance of Li | PMATFSI-PEO-PMATFSI | LFP at 70 °C at different rates from C/10 to C/2 and long-term cycling at C/2. Adapted from [213], Copyright 2017, with permission from Elsevier.

decrease the interfacial resistance between the SPE and lithium and consequently improve the battery performance. This system was tested with LFP and featured 150 mAh g^{-1} at 70 °C and C/10 for 30 cycles. These results indicate an improvement compared to the conventional PEO:LiTFSI SPE.

In general, the number of examples of SPEs in sodium batteries is still rather low compared to lithium. This is even more evident for PILs and ionomers as they have not been used as truly solid electrolytes in sodium-based batteries yet.

5.7 Alternative host material strategies

Apart from the main categories of SPE polymer hosts discussed so far, there exist a number of materials that represent alternative approaches. Common for these is generally that the ion mobility is distinctly different from the "coupled" mode of ion transport, as discussed in Chapter 2, and they are consequently generally not as strongly coordinating as many of the other polymers described in the preceding sections. These materials often also possess other appealing properties, not least in terms of mechanical strength. While a few examples are displayed in Fig. 5.49, it should also be acknowledged that some materials stemming from this category are currently undergoing commercialization in solid-state battery devices [226].

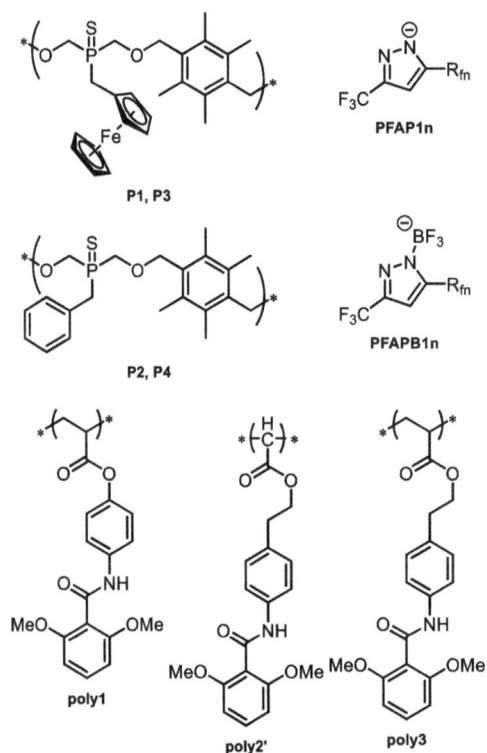

Fig. 5.49: Some SPE host materials employing alternative strategies for salt dissolution and ion transport. The perfluorinated anions PFAP1n and PFAB1n (n = 1–4) have been used togther with PVdF-HFP .

One striking example is the macromolecular platform seen in the top rows of Fig. 5.49. Through polycondensation, it is possible to synthesize a range of SPE host materials comprising copolymers between 1,4-bis(bromomethyl)-2,3,5,6-tera-methylbenzene and either ferrocenylmethyl bis(hydroxymethyl)phosphine or benzyl bis(hydroxymethyl)phosphine sulfide. These polymers, which exist as both high- and low-molecular-weight counterparts, have been referred to as **P1/P3** and **P2/P4** (Fig. 5.49), where the lower number in each pair indicates higher molecular weight [227]. These polymers display a very high T_g, in the range of 110–140 °C, which increases further when salt is added. Nevertheless, these materials are fully amorphous and up to 40 wt% LiTFSI could be dissolved into the polymers, thereby generating a 1:1 ratio between Li^+ and the oxygen atoms in the main chain. Since the associated conductivity measurements were performed far below T_g, they appeared as uncorrelated to any polymer segmental mobility. The temperature dependence of ionic conduction also did not display the typical VFT behavior. This apparent hopping mechanism, strongly resembling that in inorganic ceramic – or perhaps more correctly compared, glassy – electrolytes, could be envisioned originating in the structure, which possesses a specific combined chemosteric effect. The bulky pendant groups in combination with the main-chain phosphorus effectively hinders any packing of the polymer chains, thereby rendering a highly disordered arrangement where the apparent free volume constitutes a large fraction. A so called perpetual interconnected vacant space is thus introduced, providing ideally separated sites for the ions to jump between. While possessing a respectful conductivity, the materials also show some of the problems associated with inorganic glassy electrolytes, not least a notable brittleness.

Yet another approach to achieve an alternative transport mechanism for cations in SPEs has been to utilize the rotational motion of side groups, reminiscent of a "paddle-wheel effect" sometimes seen for inorganic ionic conductors [228]. The resulting activation energy in such systems becomes smaller, which allows for ion transport below T_g. One example of this type of SPE host material is poly[2,6-dime-thoxy-N-(4-vinylphenyl)benzamide] (called **poly1** in Fig. 5.49), where an ionic conductivity of ca. 10^{-5} S cm^{-1} has been observed for the LiTFSI-based electrolyte over the entire temperature range of –20 to 60 °C, thereby displaying a strikingly weak temperature dependence for the SPE [229]. Later, this strategy was further investigated by expanding the polymer series over **poly1, poly2'** and **poly3** [230]. Unfortunately, it showed not to be possible to reproduce the high conductivity of the SPE based on **poly1**, but **poly2'** could still be competitive as compared to PEO, particularly at low temperatures. That the more densely packed **poly2'** displays better conductivity performance than the chemically similar but less dense **poly3** supports the proposed transport mechanism, which apparently is poorly correlated to free volume. However, it should be acknowledged that while these concepts are intriguing in displaying novel mechanisms for ionic transport, they have only shown functionality at very high salt concentrations. These electrolytes possess two Li^+ per repeating polymer unit, effectively bringing it into the highly conductive but otherwise enigmatic PISE regime. Thereby the true transport mechanism is less straightforward to determine, and more profound analysis would be necessary.

Also polymers without apparent ion-coordinating abilities can sometimes display ionic conductivities, often not associated with the conventional coupled mode of transport. While this literature is extensive, it is often difficult to determine the true nature of the electrolyte material. Solvent residues can be crucial for conductivity, but are often not quantified, and the ion transport mechanism in these materials is far from fully understood. The perhaps most common such non-coordinating polymer intended for SPEs is the copolymer of vinylidene fluoride and hexafluoropropylene (PVdF-HFP), which then lacks the cation-solvating main- or side-chain oxygens or nitrogens seen in the other materials described in this chapter. For battery applications, this material has been used extensively as a binder for Li-ion battery electrodes or as a host for gel polymer electrolytes (GPEs). The high fluorination of this material endows it with a relatively high dielectric constant, which should promote ion dissociation and separation [231], despite its poor donor number – provided that the ions are in fact solvated to some degree (see Chapter 2). Thus, both PVdF-HFP:LiCF$_3$SO$_3$ [232] and PVdF-HFP:LiTFSI [233] electrolytes have been prepared and investigated. The high salt concentrations often necessary to employ for any useful conductivity, however, risk leading to salt crystallization and precipitation of solid phases in the SPEs. Recently, combined experimental and MD simulations of this system could reveal an ion transport mechanism for the lithium ions through salt channels in the amorphous regions of the non-coordinating copolymer matrix via hopping between stabilized positions at a certain percolation threshold (see Fig. 5.50) [234]. As an alternative approach but in a similar context, the concept of *fluorophilicity* can be employed to improve salt solubility in the highly fluorous PVdF-HFP matrix. Thereby, the anion–polymer interactions are utilized rather than the conventional cation–polymer coordination. For example, lithium salts based on two series of perfluorinated pyrazolide anions (PFAB1n and PFAPB1n; Fig. 5.49) in PVdF-HFP have been explored. Electronic structure calculations of these systems also indicate that both these types of anions are considerably more fluorophilic than the traditionally used CF$_3$SO$_3^-$ or TFSI anions for SPEs, although these classic SPE anions contain a significant amount of fluorine. For SPEs comprising up to 80 wt% of the LiPFAPB14 salt (thereby rendering them PISE materials), an increase in ionic conductivity with salt concentration is observed, with a maximum of 9.8×10^{-4} S cm^{-1} at 50 °C [231].

(a)

(b)

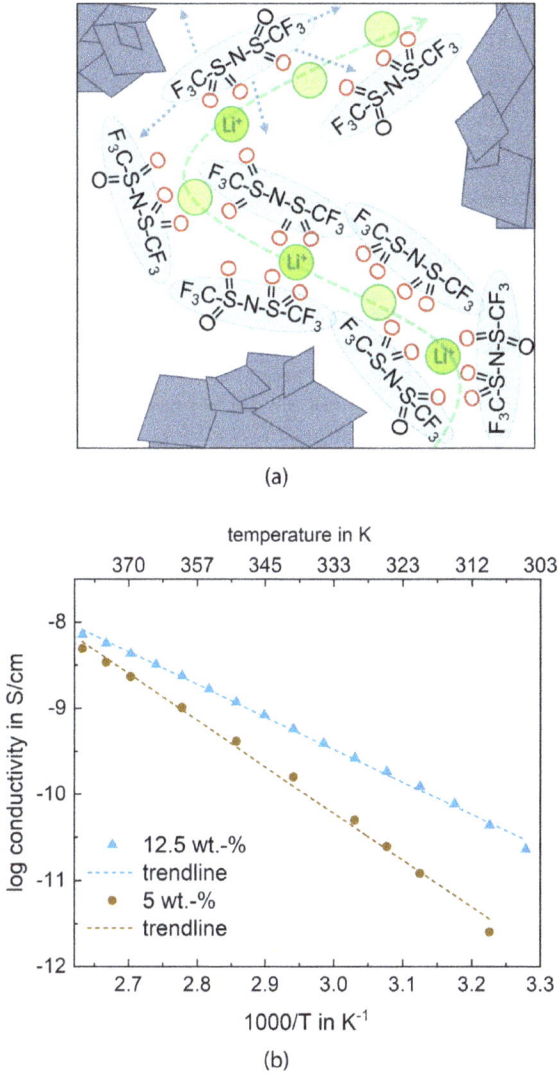

Fig. 5.50: (a) Suggested mechanism of ion transport by MD simulations in PVdF-HFP:LiTFSI, where Li$^+$ cations move via hopping in the TFSI-coordinated salt matrix forming a microstructure within the polymer and (b) corresponding Arrhenius plot for two different salt loadings. Reprinted from [234], Copyright 2020, with permission from Elsevier.

References

[1]   Harding JR, Amanchukwu CV, Hammond PT, Shao-Horn Y. Instability of Poly(ethylene oxide)
        upon oxidation in lithium–air batteries. J Phys Chem C. 2015;119:6947–55.

[2]   Herzberger J, Niederer K, Pohlit H, Seiwert J, Worm M, Wurm FR, et al. Polymerization of
        ethylene oxide, propylene oxide, and other alkylene oxides: Synthesis, novel polymer
        architectures, and bioconjugation. Chem Rev. 2016;116:2170–243.

[3]   Barteau KP, Wolffs M, Lynd NA, Fredrickson GH, Kramer EJ, Hawker CJ. Allyl glycidyl ether-
        based polymer electrolytes for room temperature lithium batteries. Macromolecules.
        2013;46:8988–94.

[4]   Alamgir M, Moulton RD, Abraham KM. Li+-conductive polymer electrolytes derived from poly
        (1,3-dioxolane) and polytetrahydrofuran. Electrochim Acta. 1991;36:773–82.

[5]   Brandrup J, Immergut EH, Grulke EA, Abe A, Bloch DR. Heats and entropies of polymerization,
        ceiling temperatures, equilibrium monomer concentrations, and polymerizability of
        heterocyclic compounds. Polymer Handbook. 4th Edition. John Wiley & Sons.

[6]   Jankowsky S, Hiller MM, Wiemhöfer HD. Preparation and electrochemical performance of
        polyphosphazene based salt-in-polymer electrolyte membranes for lithium ion batteries.
        J Power Sources. 2014;253:256–62.

[7]   Hooper R, Lyons LJ, Moline DA, West R. A highly conductive solid-state polymer electrolyte
        based on a double-comb polysiloxane polymer with Oligo(ethylene oxide) side chains.
        Organometallics. 1999;18:3249–51.

[8]   Rolland J, Brassinne J, Bourgeois JP, Poggi E, Vlad A, Gohy JF. Chemically anchored liquid-PEO
        based block copolymer electrolytes for solid-state lithium-ion batteries. J Mater Chem A.
        2014;2:11839–46.

[9]   Xue Z, He D, Xie X. Poly(ethylene oxide)-based electrolytes for lithium-ion batteries. J Mater
        Chem A. 2015;3:19218–53.

[10]   Xu K. Electrolytes and interphases in Li-Ion batteries and beyond. Chem Rev. 2014;114:
        11503–618.

[11]   Blint RJ. Binding of ether and carbonyl oxygens to lithium ion. J Electrochem Soc.
        1995;142:696–702.

[12]   Gray FM. Polymer Electrolytes. Cambridge, RSC, 1997.

[13]   Fish D, Smid J. Solvation of lithium ions in mixtures of tetraethylene glycol dimethyl ether
        and propylene carbonate. Electrochim Acta. 1992;37:2043–9.

[14]   Borodin O, Smith GD. Li+transport mechanism in Oligo(Ethylene Oxide)s compared to
        carbonates. J Solution Chem. 2007;36:803–13.

[15]   Zhang C, Ueno K, Yamazaki A, Yoshida K, Moon H, Mandai T, et al. Chelate effects in glyme/
        lithium Bis(trifluoromethanesulfonyl)amide solvate ionic liquids. i. stability of solvate cations
        and correlation with electrolyte properties. J Phys Chem B. 2014;118:5144–53.

[16]   Takahashi Y, Tadokoro H. Structural studies of polyethers, (-(CH2)m-O-)n. X. crystal structure
        of poly(ethylene oxide). Macromolecules. 1973;6:672–5.

[17]   Gadjourova Z, Andreev YG, Tunstall DP, Bruce PG. Ionic conductivity in crystalline polymer
        electrolytes. Nature. 2001;412:520–3.

[18]   Christie AM, Lilley SJ, Staunton E, Andreev YG, Bruce PG. Increasing the conductivity of
        crystalline polymer electrolytes. Nature. 2005;433:50–3.

[19]   Zhang C, Gamble S, Ainsworth D, Slawin AMZ, Andreev YG, Bruce PG. Alkali metal crystalline
        polymer electrolytes. Nat Mater. 2009;8:580–4.

[20]   Henderson WA, Passerini S. Ionic conductivity in crystalline–amorphous polymer
        electrolytes – P(EO)6: LiX phases. Electrochem Commun. 2003;5:575–8.

[21]   Liivat A, Brandell D, Thomas JO. A molecular dynamics study of ion-conduction mechanisms in crystalline low-Mw LiPF6·PEO6. J Mater Chem. 2007;17:3938–46.

[22]   Devaux D, Bouchet R, Glé D, Denoyel R. Mechanism of ion transport in PEO/LiTFSI complexes: Effect of temperature, molecular weight and end groups. Solid State Ionics. 2012;227:119–27.

[23]   Ma Q, Qi X, Tong B, Zheng Y, Feng W, Nie J, et al. Novel Li[(CF3SO2)(n-C4F9SO2)N]-Based Polymer Electrolytes for Solid-State Lithium Batteries with Superior Electrochemical Performance. ACS Appl Mater Interfaces. 2016;8:29705–12.

[24]   Shi J, Vincent CA. The effect of molecular weight on cation mobility in polymer electrolytes. Solid State Ionics. 1993;60:11–17.

[25]   Rosenwinkel MP, Andersson R, Mindemark J, Schönhoff M. Coordination effects in polymer electrolytes: Fast Li+ transport by weak ion binding. J Phys Chem C. 2020;124:23588–96.

[26]   Xu C, Sun B, Gustafsson T, Edström K, Brandell D, Hahlin M. Interface layer formation in solid polymer electrolyte lithium batteries: an XPS study. J Mater Chem A. 2014;2:7256–64.

[27]   Sun B, Xu C, Mindemark J, Gustafsson T, Edström K, Brandell D. At the polymer electrolyte interfaces: the role of the polymer host in interphase layer formation in Li-batteries. J Mater Chem A. 2015;3:13994–4000.

[28]   Xu K. Nonaqueous liquid electrolytes for lithium-based rechargeable batteries. Chem Rev. 2004;104:4303–418.

[29]   Devaux D, Glé D, Phan TNT, Gigmes D, Giroud E, Deschamps M, et al. Optimization of block copolymer electrolytes for lithium metal batteries. Chem Mater. 2015;27:4682–92.

[30]   Young W-S, Kuan W-F, Epps I, Thomas H. Block copolymer electrolytes for rechargeable lithium batteries. J Polym Sci Part B: Polym Phys. 2014;52:1–16.

[31]   Phan TN, Issa S, Gigmes D. Poly(ethylene oxide)-based block copolymer electrolytes for lithium metal batteries. Polym Int. 2019;68:7–13.

[32]   Bannister DJ, Davies GR, Ward IM, McIntyre JE. Ionic conductivities of poly(methoxy polyethylene glycol) monomethacrylate) complexes with LiSO3CH3. Polymer. 1984;25: 1600–2.

[33]   Harris DJ, Bonagamba TJ, Schmidt-Rohr K, Soo PP, Sadoway DR, Mayes AM. Solid-State NMR investigation of block copolymer electrolyte dynamics. Macromolecules. 2002;35:3772–4.

[34]   Maccallum JR, Smith MJ, Vincent CA. The effects of radiation-induced crosslinking on the conductance of LiClO4·PEO electrolytes. Solid State Ionics. 1984;11:307–12.

[35]   Lascaud S, Perrier M, Vallee A, Besner S, Prud'homme J, Armand M. Phase diagrams and conductivity behavior of Poly(ethylene oxide)-molten salt rubbery electrolytes. Macromolecules. 1994;27:7469–77.

[36]   A-vrg R, Soo PP, Sadoway DR, Mayes AM. Melt-formable block copolymer electrolytes for lithium rechargeable batteries. J Electrochem Soc. 2001;148:A537.

[37]   Gomez ED, Panday A, Feng EH, Chen V, Stone GM, Minor AM, et al. Effect of Ion Distribution on Conductivity of Block Copolymer Electrolytes. Nano Lett. 2009;9:1212–16.

[38]   Chintapalli M, Chen XC, Thelen JL, Teran AA, Wang X, Garetz BA, et al. Effect of grain size on the ionic conductivity of a block copolymer electrolyte. Macromolecules. 2014;47:5424–31.

[39]   Higa M, Yaguchi K, Kitani R. All solid-state polymer electrolytes prepared from a graft copolymer consisting of a polyimide main chain and poly(ethylene oxide) based side chains. Electrochim Acta. 2010;55:1380–4.

[40]   Li J, Lin Y, Yao H, Yuan C, Liu J. Tuning thin-film electrolyte for lithium battery by grafting cyclic carbonate and combed poly(ethylene oxide) on polysiloxane. ChemSusChem. 2014;7:1901–8.

[41]   Kunze M, Karatas Y, Wiemhöfer H-d, Eckert H, Schönhoff M. Activation of transport and local dynamics in polysiloxane-based salt-in-polymer electrolytes: A multinuclear NMR study. Phys Chem Chem Phys. 2010;12:6844–51.

[42] Zhang ZC, Jin JJ, Bautista F, Lyons LJ, Shariatzadeh N, Sherlock D, et al. Ion conductive characteristics of cross-linked network polysiloxane-based solid polymer electrolytes. Solid State Ionics. 2004;170:233–8.

[43] Blonsky PM, Shriver DF, Austin P, Allcock HR. Polyphosphazene solid electrolytes. J Am Chem Soc. 1984;106:6854–5.

[44] Hooper R, Lyons LJ, Mapes MK, Schumacher D, Moline DA, West R. Highly conductive siloxane polymers. Macromolecules. 2001;34:931–6.

[45] Devaux D, Villaluenga I, Bhatt M, Shah D, Chen XC, Thelen JL, et al. Crosslinked perfluoropolyether solid electrolytes for lithium ion transport. Solid State Ionics. 2017;310:71–80.

[46] Timachova K, Chintapalli M, Olson KR, Mecham SJ, DeSimone JM, Balsara NP. Mechanism of ion transport in perfluoropolyether electrolytes with a lithium salt. Soft Matter. 2017;13:5389–96.

[47] Wong DHC, Thelen JL, Fu Y, Devaux D, Pandya AA, Battaglia VS, et al. Nonflammable perfluoropolyether-based electrolytes for lithium batteries. Proc Natl Acad Sci USA. 2014;111:3327.

[48] Chintapalli M, Timachova K, Olson KR, Mecham SJ, Devaux D, DeSimone JM, et al. Relationship between conductivity, ion diffusion, and transference number in perfluoropolyether electrolytes. Macromolecules. 2016;49:3508–15.

[49] Cong L, Liu J, Armand M, Mauger A, Julien CM, Xie H, et al. Role of perfluoropolyether-based electrolytes in lithium metal batteries: Implication for suppressed Al current collector corrosion and the stability of Li metal/electrolytes interfaces. J Power Sources. 2018;380:115–25.

[50] Zając W, Gabryś BJ, McGreevy R, Mattssons B. Dynamics of the polymer matrix in PPO · NaClO4 and PPO · LiClO4 complexes studied by QENS. Physica B Cond Matter. 1996;226:144–51.

[51] Aldalur I, Zhang H, Piszcz M, Oteo U, Rodriguez-Martinez LM, Shanmukaraj D, et al. Jeffamine® based polymers as highly conductive polymer electrolytes and cathode binder materials for battery application. J Power Sources. 2017;347:37–46.

[52] Manning JP, Frech CB, Fung BM, Frech RE. Multinuclear nuclear magnetic resonance relaxation investigations of poly(propylene oxide) complexed with sodium trifluoromethanesulphonate. Polymer. 1991;32:2939–46.

[53] Donoso JP, Bonagamba TJ, Frare PL, Mello NC, Magon CJ, Panepucci H. Nuclear magnetic relaxation study of poly(propylene oxide)complexed with lithium salt. Electrochim Acta. 1995;40:2361–3.

[54] McLin MG, Angell CA. Frequency-dependent conductivity, relaxation times, and the conductivity/viscosity coupling problem, in polymer-electrolyte solutions: LiClO4 and NaCF3SO3 in PPO 4000. Solid State Ionics. 1992;53–56:1027–36.

[55] Watanabe M, Sanui K, Ogata N, Kobayashi T, Ohtaki Z. Ionic conductivity and mobility in network polymers from poly(propylene oxide) containing lithium perchlorate. J Appl Phys. 1985;57:123–8.

[56] Watanabe M, Nagano S, Sanui K, Ogata N. Ion conduction mechanism in network polymers from poly(ethylene oxide) and poly(propylene oxide) containing lithium perchlorate. Solid State Ionics. 1986;18–19:338–42.

[57] Ebadi M, Eriksson T, Mandal P, Costa LT, Araujo CM, Mindemark J, et al. Restricted Ion transport by plasticizing side chains in polycarbonate-based solid electrolytes. Macromolecules. 2020;53:764–74.

[58] Pesko DM, Webb MA, Jung Y, Zheng Q, Miller TF, Coates GW, et al. Universal relationship between conductivity and solvation-site connectivity in ether-based polymer electrolytes. Macromolecules. 2016;49:5244–55.

[59] Webb MA, Savoie BM, Wang Z-G, Miller Iii TF. Chemically specific dynamic bond percolation model for ion transport in polymer electrolytes. Macromolecules. 2015;48:7346–58.

[60] Mackanic DG, Michaels W, Lee M, Feng D, Lopez J, Qin J, et al. Crosslinked Poly (tetrahydrofuran) as a Loosely Coordinating Polymer Electrolyte. Adv Energy Mater. 2018;8:1800703.

[61] Akbulut O, Taniguchi I, Kumar S, Shao-Horn Y, Mayes AM. Conductivity hysteresis in polymer electrolytes incorporating poly(tetrahydrofuran). Electrochim Acta. 2007;52:1983–9.

[62] Liao YP, Liu J, Wright PV. Replies to comments contained in "Conductivity hysteresis in polymer electrolytes incorporating poly(tetrahydrofuran)" by O. Akbulut, et al., Electrochim. Acta 52 (2007) 1983. Electrochim Acta. 2007;52:7173–80.

[63] Shintani Y, Tsutsumi H. Ionic conducting behavior of solvent-free polymer electrolytes prepared from oxetane derivative with nitrile group. J Power Sources. 2010;195:2863–9.

[64] Nakano Y, Tsutsumi H. Ionic conductive properties of solid polymer electrolyte based on poly (oxetane) with branched side chains of terminal nitrile groups. Solid State Ionics. 2014;262:774–7.

[65] Nakano Y, Shinke K, Ueno K, Tsutsumi H. Solid polymer electrolytes prepared from poly (methacrylamide) derivative having tris(cyanoethoxymethyl) group as its side chain. Solid State Ionics. 2016;286:1–6.

[66] Gauthier M, Fauteux D, Vassort G, Bélanger A, Duval M, Ricoux P, et al. Assessment of Polymer-Electrolyte Batteries for EV and Ambient Temperature Applications. J Electrochem Soc. 1985;132:1333–40.

[67] Gauthier M, Fauteux D, Vassort G, Belanger A, Duval M, Ricoux P, et al. Behavior of polymer electrolyte batteries at 80 – 100 °C and near room temperature. J Power Sources. 1985;14:23–6.

[68] Kimura K, Yajima M, Tominaga Y. A highly-concentrated poly(ethylene carbonate)-based electrolyte for all-solid-state Li battery working at room temperature. Electrochem Commun. 2016;66:46–8.

[69] Trapa PE, Won -Y-Y, Mui SC, Olivetti EA, Huang B, Sadoway DR, et al. Rubbery graft copolymer electrolytes for solid-state, thin-film lithium batteries. J Electrochem Soc. 2005;152:A1.

[70] Wang F-M, Hu -C-C, Lo S-C, Wang -Y-Y, Wan -C-C. The investigation of electrochemical properties and ionic motion of functionalized copolymer electrolytes based on polysiloxane. Solid State Ionics. 2009;180:405–11.

[71] Lin Y, Li J, Lai Y, Yuan C, Cheng Y, Liu J. A wider temperature range polymer electrolyte for all-solid-state lithium ion batteries. RSC Adv. 2013;3:10722–30.

[72] Kobayashi Y, Mita Y, Seki S, Ohno Y, Miyashiro H, Terada N. Comparative study of lithium secondary batteries using nonvolatile safety electrolytes. J Electrochem Soc. 2007;154:A677.

[73] Kobayashi Y, Seki S, Mita Y, Ohno Y, Miyashiro H, Charest P, et al. High reversible capacities of graphite and SiO/graphite with solvent-free solid polymer electrolyte for lithium-ion batteries. J Power Sources. 2008;185:542–8.

[74] Fedeli E, Garcia-Calvo O, Thieu T, Phan TNT, Gigmes D, Urdampilleta I, et al. Nanocomposite solid polymer electrolytes based on semi-interpenetrating hybrid polymer networks for high performance lithium metal batteries. Electrochim Acta. 2020;353:136481.

[75] Homann G, Stolz L, Nair J, Laskovic IC, Winter M, Kasnatscheew J. Poly(Ethylene Oxide)-based Electrolyte for Solid-State-Lithium-Batteries with High Voltage Positive Electrodes: Evaluating the Role of Electrolyte Oxidation in Rapid Cell Failure. Sci Rep. 2020;10:4390.

[76] Kobayashi T, Kobayashi Y, Tabuchi M, Shono K, Ohno Y, Mita Y, et al. Oxidation reaction of polyether-based material and its suppression in lithium rechargeable battery using 4 V class cathode, LiNi1/3Mn1/3Co1/3O2. ACS Appl Mater Interfaces. 2013;5:12387–93.

[77] Doeff MM, Peng MY, Ma Y, De Jonghe LC. Orthorhombic Na x MnO2 as a cathode material for secondary sodium and lithium polymer batteries. J Electrochem Soc. 1994;141:L145–L7.

[78] Doeff MM, Ferry A, Ma Y, Ding L, De Jonghe LC. Effect of electrolyte composition on the performance of sodium/polymer cells. J Electrochem Soc. 1997;144:L20–L2.

[79] Colò F, Bella F, Nair JR, Destro M, Gerbaldi C. Cellulose-based novel hybrid polymer electrolytes for green and efficient Na-ion batteries. Electrochim Acta. 2015;174:185–90.

[80] Qi X, Ma Q, Liu L, Hu Y-S, Li H, Zhou Z, et al. Sodium bis(fluorosulfonyl)imide/poly(ethylene oxide) polymer electrolytes for sodium-ion batteries. ChemElectroChem. 2016;3:1741–5.

[81] Mindemark J, Tang S, Li H, Edman L. Ion Transport beyond the polyether paradigm: Introducing oligocarbonate ion transporters for efficient light-emitting electrochemical cells. Adv Funct Mater. 2018;28:1801295.

[82] Meabe L, Lago N, Rubatat L, Li C, Müller AJ, Sardon H, et al. Polycondensation as a versatile synthetic route to aliphatic polycarbonates for solid polymer electrolytes. Electrochim Acta. 2017;237:259–66.

[83] Rokicki G. Aliphatic cyclic carbonates and spiroorthocarbonates as monomers. Prog Polym Sci. 2000;25:259–342.

[84] Elmér AM, Jannasch P. Synthesis and characterization of poly(ethylene oxide-co-ethylene carbonate) macromonomers and their use in the preparation of crosslinked polymer electrolytes. J Polym Sci Part A: Polym Chem. 2006;44:2195–205.

[85] Tominaga Y, Shimomura T, Nakamura M. Alternating copolymers of carbon dioxide with glycidyl ethers for novel ion-conductive polymer electrolytes. Polymer. 2010;51:4295–8.

[86] Mespouille L, Coulembier O, Kawalec M, Dove AP, Dubois P. Implementation of metal-free ring-opening polymerization in the preparation of aliphatic polycarbonate materials. Prog Polym Sci. 2014;39:1144–64.

[87] Mindemark J, Törmä E, Sun B, Brandell D. Copolymers of trimethylene carbonate and ε-caprolactone as electrolytes for lithium-ion batteries. Polymer. 2015;63:91–8.

[88] Manabe Y, Uesaka M, Yoneda T, Inokuma Y. Two-Step transformation of aliphatic polyketones into π-conjugated polyimines. J Org Chem. 2019;84:9957–64.

[89] Sun B, Mindemark J,V, Morozov E, Costa LT, Bergman M, Johansson P, et al. Ion transport in polycarbonate based solid polymer electrolytes: experimental and computational investigations. Phys Chem Chem Phys. 2016;18:9504–13.

[90] Kimura K, Matsumoto H, Hassoun J, Panero S, Scrosati B, Tominaga Y. A quaternary Poly (ethylene carbonate)-Lithium Bis(trifluoromethanesulfonyl)imide-ionic liquid-silica fiber composite polymer electrolyte for lithium batteries. Electrochim Acta. 2015;175:134–40.

[91] Meabe L, Huynh TV, Lago N, Sardon H, Li C, O'Dell LA, et al. Poly(ethylene oxide carbonates) solid polymer electrolytes for lithium batteries. Electrochim Acta. 2018;264:367–75.

[92] Itoh T, Fujita K, Inoue K, Iwama H, Kondoh K, Uno T, et al. Solid polymer electrolytes based on alternating copolymers of vinyl ethers with methoxy oligo(ethyleneoxy)ethyl groups and vinylene carbonate. Electrochim Acta. 2013;112:221–9.

[93] Morioka T, Ota K, Tominaga Y. Effect of oxyethylene side chains on ion-conductive properties of polycarbonate-based electrolytes. Polymer. 2016;84:21–6.

[94] Tominaga Y, Nanthana V, Tohyama D. Ionic conduction in poly(ethylene carbonate)-based rubbery electrolytes including lithium salts. Polym J. 2012;44:1155–8.

[95] Tominaga Y, Yamazaki K. Fast Li-ion conduction in poly(ethylene carbonate)-based electrolytes and composites filled with TiO2 nanoparticles. Chem Commun. 2014;50:4448–50.

[96] Morioka T, Nakano K, Tominaga Y. Ion-conductive properties of a polymer electrolyte based on ethylene carbonate/ethylene oxide random copolymer. Macromol Rapid Commun. 2017;38:1600652.

[97] Kimura K, Motomatsu J, Tominaga Y. Highly concentrated polycarbonate-based solid polymer electrolytes having extraordinary electrochemical stability. J Polym Sci Part B: Polym Phys. 2016;54:2442–7.

[98] Zhang J, Zhao J, Yue L, Wang Q, Chai J, Liu Z, et al. Safety-reinforced poly(propylene carbonate)-based all-solid-state polymer electrolyte for ambient-temperature solid polymer lithium batteries. Adv Energy Mater. 2015;5:1501082.

[99] Li Y, Ding F, Xu Z, Sang L, Ren L, Ni W, et al. Ambient temperature solid-state Li-battery based on high-salt-concentrated solid polymeric electrolyte. J Power Sources. 2018;397:95–101.

[100] Fei H, Liu Y, An Y, Xu X, Zeng G, Tian Y, et al. Stable all-solid-state potassium battery operating at room temperature with a composite polymer electrolyte and a sustainable organic cathode. J Power Sources. 2018;399:294–8.

[101] Wang C, Zhang H, Li J, Chai J, Dong S, Cui G. The interfacial evolution between polycarbonate-based polymer electrolyte and Li-metal anode. J Power Sources. 2018;397:157–61.

[102] Commarieu B, Paolella A, Collin-Martin S, Gagnon C, Vijh A, Guerfi A, et al. Solid-to-liquid transition of polycarbonate solid electrolytes in Li-metal batteries. J Power Sources. 2019;436:226852.

[103] Melchiors M, Keul H, Höcker H. Preparation and properties of solid electrolytes on the basis of alkali metal salts and poly(2,2-dimethyltrimethylene carbonate)-block-poly(ethylene oxide)-block-poly(2,2-dimethyltrimethylene carbonate). Polymer. 1996;37:1519–27.

[104] Smith MJ, Silva MM, Cerqueira S, MacCallum JR. Preparation and characterization of a lithium ion conducting electrolyte based on poly(trimethylene carbonate). Solid State Ionics. 2001;140:345–51.

[105] Sun B, Mindemark J, Edström K, Brandell D. Polycarbonate-based solid polymer electrolytes for Li-ion batteries. Solid State Ionics. 2014;262:738–42.

[106] Mindemark J, Mogensen R, Smith MJ, Silva MM, Brandell D. Polycarbonates as alternative electrolyte host materials for solid-state sodium batteries. Electrochem Commun. 2017;77:58–61.

[107] Sångeland C, Mogensen R, Brandell D, Mindemark J. Stable cycling of sodium metal all-solid-state batteries with polycarbonate-based polymer electrolytes. ACS Appl Polym Mater. 2019;1:825–32.

[108] Mindemark J, Imholt L, Brandell D. Synthesis of high molecular flexibility polycarbonates for solid polymer electrolytes. Electrochim Acta. 2015;175:247–53.

[109] Mindemark J, Sun B, Brandell D. Hydroxyl-functionalized poly(trimethylene carbonate) electrolytes for 3D-electrode configurations. Polym Chem. 2015;6:4766–74.

[110] He W, Cui Z, Liu X, Cui Y, Chai J, Zhou X, et al. Carbonate-linked poly(ethylene oxide) polymer electrolytes towards high performance solid state lithium batteries. Electrochim Acta. 2017;225:151–9.

[111] Forsyth M, Tipton AL, Shriver DF, Ratner MA, MacFarlane DR. Ionic conductivity in poly (diethylene glycol-carbonate)/sodium triflate complexes. Solid State Ionics. 1997;99:257–61.

[112] Wei X, Shriver DF. Highly conductive polymer electrolytes containing rigid polymers. Chem Mater. 1998;10:2307–8.

[113] Wang Y, Fan F, Agapov AL, Saito T, Yang J, Yu X, et al. Examination of the fundamental relation between ionic transport and segmental relaxation in polymer electrolytes. Polymer. 2014;55:4067–76.

[114] Watanabe M, Rikukawa M, Sanui K, Ogata N, Kato H, Kobayashi T, et al. Ionic conductivity of polymer complexes formed by poly(ethylene succinate) and lithium perchlorate. Macromolecules. 1984;17:2902–8.

[115] Wu ID, Chang F-C. Determination of the interaction within polyester-based solid polymer electrolyte using FTIR spectroscopy. Polymer. 2007;48:989–96.

[116] Lin C-K, Wu ID. Investigating the effect of interaction behavior on the ionic conductivity of Polyester/LiClO4 blend systems. Polymer. 2011;52:4106–13.

[117] Watanabe M, Rikukawa M, Sanui K, Ogata N. Effects of polymer structure and incorporated salt species on ionic conductivity of polymer complexes formed by aliphatic polyester and alkali metal thiocyanate. Macromolecules. 1986;19:188–92.

[118] Dupon R, Papke BL, Ratner MA, Shriver DF. Ion transport in the polymer electrolytes formed between Poly(ethylene succinate) and lithium tetrafluoroborate. J Electrochem Soc. 1984;131:586–9.

[119] Lee Y-C, Ratner MA, Shriver DF. Ionic conductivity in the poly(ethylene malonate)/lithium triflate system. Solid State Ionics. 2001;138:273–6.

[120] Mindemark J, Sun B, Törmä E, Brandell D. High-performance solid polymer electrolytes for lithium batteries operational at ambient temperature. J Power Sources. 2015;298:166–70.

[121] Polo Fonseca C, Neves S. Electrochemical properties of a biodegradable polymer electrolyte applied to a rechargeable lithium battery. J Power Sources. 2006;159:712–6.

[122] Fonseca CP, Rosa DS, Gaboardi F, Neves S. Development of a biodegradable polymer electrolyte for rechargeable batteries. J Power Sources. 2006;155:381–4.

[123] Sångeland C, Younesi R, Mindemark J, Brandell D. Towards room temperature operation of all-solid-state Na-ion batteries through polyester–polycarbonate-based polymer electrolytes. Energy Storage Mater. 2019;19:31–8.

[124] Eriksson T, Mindemark J, Yue M, Brandell D. Effects of nanoparticle addition to poly(ε-caprolactone) electrolytes: Crystallinity, conductivity and ambient temperature battery cycling. Electrochim Acta. 2019;300:489–96.

[125] Florjańczyk Z, Zygadło-Monikowska E, Wieczorek W, Ryszawy A, Tomaszewska A, Fredman K, et al. Polymer-in-salt electrolytes based on acrylonitrile/butyl acrylate copolymers and lithium salts. J Phys Chem B. 2004;108:14907–14.

[126] Florjańczyk Z, Zygadło-Monikowska E, Affek A, Tomaszewska A, Łasińska A, Marzantowicz M, et al. Polymer electrolytes based on acrylonitrile–butyl acrylate copolymers and lithium bis (trifluoromethanesulfone)imide. Solid State Ionics. 2005;176:2123–8.

[127] Łasińska AK, Marzantowicz M, Dygas JR, Krok F, Florjańczyk Z, Tomaszewska A, et al. Study of ageing effects in polymer-in-salt electrolytes based on poly(acrylonitrile-co-butyl acrylate) and lithium salts. Electrochim Acta. 2015;169:61–72.

[128] Eriksson T, Mace A, Manabe Y, Yoshizawa-Fujita M, Inokuma Y, Brandell D, et al. Polyketones as host materials for solid polymer electrolytes. J Electrochem Soc. 2020;167:070537.

[129] Sun B, Mindemark J, Edström K, Brandell D. Realization of high performance polycarbonate-based Li polymer batteries. Electrochem Commun. 2015;52:71–4.

[130] Chai J, Liu Z, Ma J, Wang J, Liu X, Liu H, et al. In situ generation of Poly (Vinylene Carbonate) based solid electrolyte with interfacial stability for LiCoO2 lithium batteries. Adv Sci. 2017;4:1600377.

[131] Sun B, Asfaw HD, Rehnlund D, Mindemark J, Nyholm L, Edström K, et al. Toward solid-state 3D-microbatteries using functionalized polycarbonate-based polymer electrolytes. ACS Appl Mater Interfaces. 2018;10:2407–13.

[132] Isken P, Dippel C, Schmitz R, Schmitz RW, Kunze M, Passerini S, et al. High flash point electrolyte for use in lithium-ion batteries. Electrochim Acta. 2011;56:7530–5.

[133] Yamada Y, Furukawa K, Sodeyama K, Kikuchi K, Yaegashi M, Tateyama Y, et al. Unusual stability of acetonitrile-based superconcentrated electrolytes for fast-charging lithium-ion batteries. J Am Chem Soc. 2014;136:5039–46.

[134] Long S, MacFarlane DR, Forsyth M. Fast ion conduction in molecular plastic crystals. Solid State Ionics. 2003;161:105–12.

[135] Fan L-Z, Hu Y-S, Bhattacharyya AJ, Maier J. Succinonitrile as a Versatile Additive for Polymer Electrolytes. Adv Funct Mater. 2007;17:2800–7.

[136] He R, Echeverri M, Ward D, Zhu Y, Kyu T. Highly conductive solvent-free polymer electrolyte membrane for lithium-ion batteries: Effect of prepolymer molecular weight. J Membr Sci. 2016;498:208–17.

[137] Appetecchi GB, Croce F, Romagnoli P, Scrosati B, Heider U, Oesten R. High-performance gel-type lithium electrolyte membranes. Electrochem Commun. 1999;1:83–6.

[138] Hu P, Chai J, Duan Y, Liu Z, Cui G, Chen L. Progress in nitrile-based polymer electrolytes for high performance lithium batteries. J Mater Chem A. 2016;4:10070–83.

[139] Chen-Yang YW, Chen HC, Lin FJ, Chen CC. Polyacrylonitrile electrolytes: 1. A novel high-conductivity composite polymer electrolyte based on PAN, LiClO4 and α-Al2O3. Solid State Ionics. 2002;150:327–35.

[140] Voigt N, van Wüllen L. The mechanism of ionic transport in PAN-based solid polymer electrolytes. Solid State Ionics. 2012;208:8–16.

[141] Watanabe M, Kanba M, Nagaoka K, Shinohara I. Ionic conductivity of hybrid films composed of polyacrylonitrile, ethylene carbonate, and LiClO4. J Polym Sci: Polym Phys Ed. 1983;21:939–48.

[142] Sim LN, Sentanin FC, Pawlicka A, Yahya R, Arof AK. Development of polyacrylonitrile-based polymer electrolytes incorporated with lithium bis(trifluoromethane)sulfonimide for application in electrochromic device. Electrochim Acta. 2017;229:22–30.

[143] Forsyth M, Sun JZ, MacFarlane DR. Novel polymer-in-salt electrolytes based on polyacrylonitrile (PAN) lithium triflate salt mixtures. Solid State Ionics. 1998;112:161–3.

[144] Forsyth M, Sun J, Macfarlane DR, Hill AJ. Compositional dependence of free volume in PAN/LiCF3SO3 polymer-in-salt electrolytes and the effect on ionic conductivity. J Polym Sci Part B: Polym Phys. 2000;38:341–50.

[145] Ferry A, Edman L, Forsyth M, MacFarlane DR, Sun J. Connectivity, ionic interactions, and migration in a fast-ion-conducting polymer-in-salt electrolyte based on poly(acrylonitrile) and LiCF3SO3. J Appl Phys. 1999;86:2346–8.

[146] Huang Z-H, Tsai D-S, Chiu C-J, Pham Q-T, Chern C-S. A lithium solid electrolyte of acrylonitrile copolymer with thiocarbonate moiety and its potential battery application. Electrochim Acta. 2021;365:137357.

[147] Chen YT, Chuang YC, Su JH, Yu HC, Chen-Yang YW. High discharge capacity solid composite polymer electrolyte lithium battery. J Power Sources. 2011;196:2802–9.

[148] Saunier J, Alloin F, Sanchez JY. Electrochemical and spectroscopic studies of polymethacrylonitrile based electrolytes. Electrochim Acta. 2000;45:1255–63.

[149] Bushkova OV, Animitsa IE, Lirova BI, Zhukovsky VM. Lithium conducting solid polymer electrolytes based on polyacrylonitrile copolymers: Ion solvation and transport properties. Ionics. 1997;3:396–404.

[150] Bushkova OV, Zhukovsky VM, Lirova BI, Kruglyashov AL. Fast ionic transport in solid polymer electrolytes based on acrylonitrile copolymers. Solid State Ionics. 1999;119:217–22.

[151] Bushkova OV, Popov SE, Yaroslavtseva TV, Zhukovsky VM, Nikiforov AE. Ion–molecular and ion–ion interactions in solvent-free polymer electrolytes based on amorphous butadiene – acrylontrile copolymer and LiAsF6. Solid State Ionics. 2008;178:1817–30.

[152] Yaroslavtseva TV, Bushkova OV. Glass transitions and ionic conductivity in a poly(butadiene-acrylonitrile)–LiAsF6 system. Electrochim Acta. 2011;57:212–9.

[153] Erickson M, Frech R, Glatzhofer DT. Solid polymer/salt electrolytes based on linear poly((N-2-cyanoethyl)ethylenimine). Electrochim Acta. 2003;48:2059–63.

[154] Tanaka R, Fujita T, Nishibayashi H, Saito S. Ionic conduction in poly(ethylenimine)-and poly(N-methylethylenimine)-lithium salt systems. Solid State Ionics. 1993;60:119–23.

[155] Ionescu-Vasii LL, Garcia B, Armand M. Conductivities of electrolytes based on PEI-b-PEO-b-PEI triblock copolymers with lithium and copper TFSI salts. Solid State Ionics. 2006;177:885–92.

[156] Chiang CK, Davis GT, Harding CA, Takahashi T. Polymeric electrolyte based on poly(ethylene imine) and lithium salts. Solid State Ionics. 1986;18–19:300–5.

[157] Tanaka R, Ueoka I, Takaki Y, Kataoka K, Saito S. High molecular weight linear polyethylenimine and poly(N-methylethylenimine). Macromolecules. 1983;16:849–53.

[158] Saegusa T, Ikeda H, Fujii H. Crystalline polyethylenimine. Macromolecules. 1972;5:108-.

[159] Saegusa T, Ikeda H, Fujii H. Isomerization polymerization of 2-oxazoline. I. preparation of unsubstituted 2-oxazoline polymer. Polym J. 1972;3:35–9.

[160] Jones GD, MacWilliams DC, Braxtor NA. Species in the Polymerization of Ethylenimine and N-Methylethylenimine. J Org Chem. 1965;30:1994–2003.

[161] Paul JL, Jegat C, Lassègues JC. Branched poly(ethyleneimine)-CF3SO3Li complexes. Electrochim Acta. 1992;37:1623–5.

[162] Chiang CK, Davis GT, Harding CA, Takahashi T. Polyethylenimine-sodium iodide complexes. Macromolecules. 1985;18:825–7.

[163] York S, Frech R, Snow A, Glatzhofer D. A comparative vibrational spectroscopic study of lithium triflate and sodium triflate in linear poly(ethylenimine). Electrochim Acta. 2001;46:1533–7.

[164] York SS, Buckner M, Frech R. Ion–Polymer and Ion–Ion Interactions in Linear Poly (ethylenimine) Complexed with LiCF3SO3 and LiSbF6. Macromolecules. 2004;37:994–9.

[165] Tanaka R, Sakurai M, Sekiguchi H, Mori H, Murayama T, Ooyama T. Lithium ion conductivity in polyoxyethylene/polyethylenimine blends. Electrochim Acta. 2001;46:1709–15.

[166] Tanaka R, Sakurai M, Sekiguchi H, Inoue M. Improvement of room-temperature conductivity and thermal stability of PEO–LiClO4 systems by addition of a small proportion of polyethylenimine. Electrochim Acta. 2003;48:2311–6.

[167] Pehlivan İB, Marsal R, Niklasson GA, Granqvist CG, Georén P. PEI–LiTFSI electrolytes for electrochromic devices: Characterization by differential scanning calorimetry and viscosity measurements. Sol Energy Mater Sol Cells. 2010;94:2399–404.

[168] Pehlivan İB, Georén P, Marsal R, Granqvist CG, Niklasson GA. Ion conduction of branched polyethyleneimine–lithium bis(trifluoromethylsulfonyl) imide electrolytes. Electrochim Acta. 2011;57:201–6.

[169] Hu L, Frech R, Glatzhofer DT, Mason R, York SS. Linear poly(propylenimine)/lithium triflate as a polymer electrolyte system. Solid State Ionics. 2008;179:401–8.

[170] Sanoja GE, Schauser NS, Bartels JM, Evans CM, Helgeson ME, Seshadri R, et al. Ion transport in dynamic polymer networks based on metal–ligand coordination: Effect of cross-linker concentration. Macromolecules. 2018;51:2017–26.

[171] Schauser NS, Sanoja GE, Bartels JM, Jain SK, Hu JG, Han S, et al. Decoupling bulk mechanics and mono- and multivalent ion transport in polymers based on metal–ligand coordination. Chem Mater. 2018;30:5759–69.

[172] Kanbara T, Inami M, Yamamoto T, Nishikata A, Tsuru T, Watanabe M, et al. New lithium salt ionic conductor using Poly(vinyl alcohol) matrix. Chem Lett. 1989;18:1913–16.

[173] Friis N, Goosney D, Wright JD, Hamielec AE. Molecular weight and branching development in vinyl acetate emulsion polymerization. J Appl Polym Sci. 1974;18:1247–59.

[174] Minsk LM, Priest WJ, Kenyon WO. The alcoholysis of polyvinyl acetate*. J Am Chem Soc. 1941;63:2715–21.

[175] Bunn CW. Crystal structure of polyvinyl alcohol. Nature. 1948;161:929–30.

[176] MacFarlane DR, Zhou F, Forsyth M. Ion conductivity in amorphous polymer/salt mixtures. Solid State Ionics. 1998;113–115:193–7.

[177] Ek G, Jeschull F, Bowden T, Brandell D. Li-ion batteries using electrolytes based on mixtures of poly(vinyl alcohol) and lithium bis(triflouromethane) sulfonamide salt. Electrochim Acta. 2017;246:208–12.

[178] Every HA, Zhou F, Forsyth M, MacFarlane DR. Lithium ion mobility in poly(vinyl alcohol) based polymer electrolytes as determined by 7Li NMR spectroscopy. Electrochim Acta. 1998;43:1465–9.

[179] Noor IS, Majid SR, Arof AK. Poly(vinyl alcohol)–LiBOB complexes for lithium–air cells. Electrochim Acta. 2013;102:149–60.

[180] Park H-K, Kong B-S, Oh E-S. Effect of high adhesive polyvinyl alcohol binder on the anodes of lithium ion batteries. Electrochem Commun. 2011;13:1051–3.

[181] Song J, Zhou M, Yi R, Xu T, Gordin ML, Tang D, et al. Interpenetrated gel polymer binder for high-performance silicon anodes in lithium-ion batteries. Adv Funct Mater. 2014;24:5904–10.

[182] Capek I. Nature and properties of ionomer assemblies II.. Adv Colloid Interface Sci. 2005;118:73–112.

[183] Middleton LR, Winey KI. Nanoscale aggregation in acid- and ion-containing polymers. Annu Rev Chem Biomol Eng. 2017;8:499–523.

[184] Yuan J, Antonietti M. Poly(ionic liquid)s: Polymers expanding classical property profiles. Polymer. 2011;52:1469–82.

[185] Mecerreyes D. Polymeric ionic liquids: Broadening the properties and applications of polyelectrolytes. Prog Polym Sci. 2011;36:1629–48.

[186] Ohno H, Yoshizawa M, Ogihara W. Development of new class of ion conductive polymers based on ionic liquids. Electrochim Acta. 2004;50:255–61.

[187] Hirao M, Ito K, Ohno H. Preparation and polymerization of new organic molten salts; N-alkylimidazolium salt derivatives. Electrochim Acta. 2000;45:1291–4.

[188] Ohno H, Ito K. Room-temperature molten salt polymers as a matrix for fast ion conduction. Chem Lett. 1998;27:751–2.

[189] Washiro S, Yoshizawa M, Nakajima H, Ohno H. Highly ion conductive flexible films composed of network polymers based on polymerizable ionic liquids. Polymer. 2004;45:1577–82.

[190] Tant MR, Wilkes GL. An overview of the viscous and viscoelastic behavior of ionomers in bulk and solution. J Macromol Sci Part C. 1988;28:1–63.

[191] Capek I. Dispersions of polymer ionomers: I. Adv Colloid Interface Sci. 2004;112:1–29.

[192] Chen Q, Bao N, Wang J-H-H, Tunic T, Liang S, Colby RH, Viscoelasticity L. Dielectric spectroscopy of ionomer/plasticizer mixtures: A transition from ionomer to polyelectrolyte. Macromolecules. 2015;48:8240–52.

[193] Watkin RR Ionic hydrocarbon polymers. 1961.

[194] Bouchet R, Maria S, Meziane R, Aboulaich A, Lienafa L, Bonnet J-P, et al. Single-ion BAB triblock copolymers as highly efficient electrolytes for lithium-metal batteries. Nat Mater. 2013;12:452–7.

[195] Armand M. The history of polymer electrolytes. Solid State Ionics. 1994;69:309–19.

[196] Jangu C, Savage AM, Zhang Z, Schultz AR, Madsen LA, Beyer FL, et al. Sulfonimide-containing triblock copolymers for improved conductivity and mechanical performance. Macromolecules. 2015;48:4520–8.

[197] Ma Q, Zhang H, Zhou C, Zheng L, Cheng P, Nie J, et al. Single Lithium-Ion Conducting Polymer Electrolytes Based on a Super-Delocalized Polyanion. Angew Chem Int Ed. 2016;55:2521–5.

[198] Li J, Zhu H, Wang X, Armand M, MacFarlane DR, Forsyth M. Synthesis of sodium Poly[4-styrenesulfonyl(trifluoromethylsulfonyl)imide]-co-ethylacrylate] solid polymer electrolytes. Electrochim Acta. 2015;175:232–9.

[199] Mohd Noor SA, Gunzelmann D, Sun J, MacFarlane DR, Forsyth M. Ion conduction and phase morphology in sulfonate copolymer ionomers based on ionic liquid–sodium cation mixtures. J Mater Chem A. 2014;2:365–74.

[200] Rolland J, Poggi E, Vlad A, Gohy J-F. Single-ion diblock copolymers for solid-state polymer electrolytes. Polymer. 2015;68:344–52.

[201] Yan L, Hoang L, Winey KI. Ionomers from step-growth polymerization: Highly ordered ionic aggregates and ion conduction. Macromolecules. 2020;53:1777–84.

[202] Wang S-W, Colby RH, Viscoelasticity L. Cation conduction in polyurethane sulfonate ionomers with ions in the soft segment–multiphase systems. Macromolecules. 2018;51:2767–75.

[203] Dou S, Zhang S, Klein RJ, Runt J, Colby RH. Synthesis and characterization of poly(Ethylene Glycol)-based single-ion conductors. Chem Mater. 2006;18:4288–95.

[204] Hunley MT, England JP, Long TE. Influence of counteranion on the thermal and solution behavior of Poly(2-(dimethylamino)ethyl methacrylate)-based polyelectrolytes. Macromolecules. 2010;43:9998–10005.

[205] Wang W, Liu W, Tudryn GJ, Colby RH, Winey KI. Multi-length scale morphology of Poly (ethylene oxide)-based sulfonate ionomers with alkali cations at room temperature. Macromolecules. 2010;43:4223–9.

[206] Delhorbe V, Bresser D, Mendil-Jakani H, Rannou P, Bernard L, Gutel T, et al. Unveiling the ion conduction mechanism in imidazolium-based Poly(ionic liquids): A comprehensive investigation of the structure-to-transport interplay. Macromolecules. 2017;50:4309–21.

[207] Klein RJ, Welna DT, Weikel AL, Allcock HR, Runt J. Counterion effects on ion mobility and mobile ion concentration of doped polyphosphazene and polyphosphazene ionomers. Macromolecules. 2007;40:3990–5.

[208] Wang S-W, Liu W, Colby RH. Counterion dynamics in polyurethane-carboxylate ionomers with ionic liquid counterions. Chem Mater. 2011;23:1862–73.

[209] Wang S-W, Colby RH. Linear viscoelasticity and cation conduction in polyurethane sulfonate ionomers with ions in the soft segment–single phase systems. Macromolecules. 2018;51:2757–66.

[210] Inceoglu S, Rojas AA, Devaux D, Chen XC, Stone GM, Balsara NP. Morphology–conductivity relationship of single-ion-conducting block copolymer electrolytes for lithium batteries. ACS Macro Lett. 2014;3:510–4.

[211] Kasemägi H, Ollikainen M, Brandell D, Aabloo A. Molecular dynamics modelling of block-copolymer electrolytes with high t+ values. Electrochim Acta. 2015;175:47–54.

[212] Devaux D, Liénafa L, Beaudoin E, Maria S, Phan TNT, Gigmes D, et al. Comparison of single-ion-conductor block-copolymer electrolytes with Polystyrene-TFSI and Polymethacrylate-TFSI structural blocks. Electrochim Acta. 2018;269:250–61.

[213] Porcarelli L, Aboudzadeh MA, Rubatat L, Nair JR, Shaplov AS, Gerbaldi C, et al. Single-ion triblock copolymer electrolytes based on poly(ethylene oxide) and methacrylic sulfonamide blocks for lithium metal batteries. J Power Sources. 2017;364:191–9.

[214] Porcarelli L, Vlasov PS, Ponkratov DO, Lozinskaya EI, Antonov DY, Nair JR, et al. Design of ionic liquid like monomers towards easy-accessible single-ion conducting polymer electrolytes. Eur Polym J. 2018;107:218–28.

[215] Salas-de La Cruz D, Green MD, Ye Y, Ya E, Long TE, Winey KI. Correlating backbone-to-backbone distance to ionic conductivity in amorphous polymerized ionic liquids. J Polym Sci Part B: Polym Phys. 2012;50:338–46.

[216] Sampath J, Hall LM. Impact of ionic aggregate structure on ionomer mechanical properties from coarse-grained molecular dynamics simulations. J Chem Phys. 2017;147:134901.

[217] Frischknecht AL, Paren BA, Middleton LR, Koski JP, Tarver JD, Tyagi M, et al. Chain and ion dynamics in precise polyethylene ionomers. Macromolecules. 2019;52:7939–50.

[218] Chen X, Forsyth M, Chen F. Molecular dynamics study of ammonium based co-cation plasticizer effect on lithium ion dynamics in ionomer electrolytes. Solid State Ionics. 2018;316:47–52.

[219] Tudryn GJ, Liu W, Wang S-W, Colby RH. Counterion dynamics in polyester–sulfonate ionomers with ionic liquid counterions. Macromolecules. 2011;44:3572–82.

[220] Li J, Zhu H, Wang X, MacFarlane DR, Armand M, Forsyth M. Increased ion conduction in dual cation [sodium][tetraalkylammonium] poly[4-styrenesulfonyl(trifluoromethylsulfonyl)imide-co-ethylacrylate] ionomers. J Mater Chem A. 2015;3:19989–95.

[221] Piszcz M, Garcia-Calvo O, Oteo U, Lopez Del Amo JM, Li C, Lm R-M, et al. New single ion conducting blend based on PEO and PA-LiTFSI. Electrochim Acta. 2017;255:48–54.

[222] Lozinskaya EI, Cotessat M, Shmalko AV, Ponkratov DO, Gumileva LV, Sivaev IB, et al. Expanding the chemistry of single-ion conducting poly(ionic liquid)s with polyhedral boron anions. Polym Int. 2019;68:1570–9.

[223] Martinez-Ibañez M, Sanchez-Diez E, Qiao L, Meabe L, Santiago A, Zhu H, et al. Weakly coordinating fluorine-free polysalt for single lithium-ion conductive solid polymer electrolytes. Batteries & Supercaps. 2020;3:738–46.

[224] Hallinan DT, Balsara NP. Polymer electrolytes. Annu Rev Mater Res. 2013;43:503–25.

[225] Porcarelli L, Shaplov AS, Salsamendi M, Nair JR, Vygodskii YS, Mecerreyes D, et al. Single-Ion block Copoly(ionic liquid)s as electrolytes for all-solid state lithium batteries. ACS Appl Mater Interfaces. 2016;8:10350–9.

[226] Wood KN, Noked M, Dasgupta NP. Lithium Metal Anodes: Toward an Improved Understanding of Coupled Morphological, Electrochemical, and Mechanical Behavior. ACS Energy Lett 2017,2,664–72.

[227] Babu HV, Srinivas B, Muralidharan K. Design of polymers with an intrinsic disordered framework for Li-ion conducting solid polymer electrolytes. Polymer. 2015;75:10–16.

[228] Kaber R, Nilsson L, Andersen NH, Lunden A, Thomas JO. A single-crystal neutron diffraction study of the structure of the high-temperature rotor phase of lithium sulphate. J Phys Condens Matter. 1992;4:1925–33.

[229] Sato A, Okumura T, Nishimura S, Yamamoto H, Ueyama N. Lithium ion conductive polymer electrolyte by side group rotation. J Power Sources. 2005;146:423–6.

[230] Shimomoto H, Uegaito T, Yabuki S, Teratani S, Itoh T, Ihara E, et al. Lithium ion conductivity of polymers containing N-phenyl-2,6-dimethoxybenzamide framework in their side chains: Possible role of bond rotation in polymer side chain substituents for efficient ion transport. Solid State Ionics. 2016;292:1–7.

[231] Cznotka E, Jeschke S, Grünebaum M, Wiemhöfer H-D. Highly-fluorous pyrazolide-based lithium salt in PVDF-HFP as solid polymer electrolyte. Solid State Ionics. 2016;292:45–51.

[232] Ramesh S, Lu S-C. Effect of lithium salt concentration on crystallinity of poly(vinylidene fluoride-co-hexafluoropropylene)-based solid polymer electrolytes. J Mol Struct. 2011;994:403–9.

[233] Stephan AM, Nahm KS, Anbu Kulandainathan M, Ravi G, Wilson J. Poly(vinylidene fluoride-hexafluoropropylene) (PVdF-HFP) based composite electrolytes for lithium batteries. Eur Polym J. 2006;42:1728–34.

[234] Mathies L, Diddens D, Dong D, Bedrov D, Leipner H. Transport mechanism of lithium ions in non-coordinating P(VdF-HFP) copolymer matrix. Solid State Ionics. 2020;357:115497.

6 Outlook

The overview of different polymer host materials given in Chapter 5 gives some guide-lines for the potential application of different SPEs in batteries, especially if consider-ing the requirements discussed in Chapter 4. It is, however, striking that there seems not yet to be a single homopolymer that can intrinsically meet all challenges put for-ward for implementation in high-energy-density batteries – these SPEs generally fail in terms of either mechanical integrity or flexibility, anodic or reductive stability, ionic conductivity or cation transport numbers. On the other hand, considering the general versatility of polymer chemistry and the possibility to combine several appealing prop-erties into one material – being it a homogeneous blend, an interpenetrating polymer network, bicontinuous phases, a layer-by-layer architecture, a block copolymer, etc. – it is not far-fetched to imagine straightforward fabrication of an SPE that functions well with a metal electrode, can infiltrate a porous cathode, provides mechanical sepa-ration and shows decent cation conductivity in an appropriate temperature interval. However, no such SPE exists as of yet. From the main categories of polymer electrolyte hosts summarized in Chapter 5, it has for example been frequently shown that poly-ethers display compatibility with metal electrodes and that carbonyl-coordinating mo-tifs or single-ion polymers show high Li^+ transport numbers. There are also indications that polynitriles feature stability with high-voltage electrodes, while many hosts pos-sessing a decoupled mode of transport can provide mechanisms for high conductivity. Adding cross-linking and block-copolymeric strategies with robust segments can then help tailoring the mechanical properties. The challenge is thus to combine these ap-pealing properties of different parts, while avoiding to create a Frankenstein's monster.

In this context, one straightforward idea that has been approached recently is to construct double-layer polymer electrolytes – one SPE layer that is adopted for the anode (generally a polyether) and another for the cathode [1]. One immediate concern then appears: that yet another interface is introduced in the battery cell with its own interfacial chemistry and associated resistance. However, considering that both SPE layers are soft materials, it should be able to chemically modify this interface to facilitate ionic transport across it. Fabrication of advanced polymer ma-terials has historically overcome much larger challenges. The materials develop-ment of such multicomponent SPE systems has merely begun.

Likewise striking – and somewhat surprising – when summarizing the data of dif-ferent SPEs in Chapter 5 is that T_g seems to be less of a crucial factor than what the conventional theory of ionic transport described in Chapter 2 stresses. The correlation between T_g and ionic conductivity is not always very strong, and the correlation with battery behavior is even weaker. Generally, both high- and low-T_g ion-coordinating pol-ymers display some ionic conductivity, and not rarely with rather similar values. This highlights the importance of other mobility phenomena than those strongly linked to polymer segmental motion. On the other hand, as seen for PAN- and PVA-based

https://doi.org/10.1515/9781501521140-006

materials, the reported conductivity in some high-T_g systems seems to also be strongly coupled to the remaining solvent residues, often in quite large amounts. As pointed out previously in this book, while this can lead to materials displaying good conductivity performance and short-term useful battery cycling data, it is largely unclear how these batteries will age. Moreover, if the SPE system becomes more complex by incorporation of several different components, remaining liquid components might interact with these incorporated species in an unpredictable manner. Some recent research has more strongly focused on solvent residues and hygroscopicity in SPE materials [2], which certainly is welcome to better understand the applicability of different materials and elucidate their ion transport mechanisms. It should, however, not be ruled out that confined liquids in the SPE matrix might be a very useful addition in terms of improving conductivity properties – if such liquid conductivity enhancers are chemically and electrochemically stable, they can provide game-changing properties for the electrolyte systems. The problem is then to monitor and control, and ultimately tailor, this stability.

Nevertheless, considering the weak inverse correlation seen for many SPE systems between conductivity and T_g, we can foresee a continued development of polymeric systems with comparatively high T_g. Such materials will more easily retain their integrity at high operating temperatures, thereby being less dependent on cross-linking strategies or external separators. Chasing materials with increasingly low T_g would bring the SPE down to the near-liquid domain if useful ionic conductivity values are to be obtained [3] – at least for a homopolymer system, where good room-temperature battery performance remains an elusive dream. The improved conductivity seen for not least the "alternative" approaches (Section 5.7) often stems from a "structurization" of the polymer host, providing useful paths for ionic migration. This development toward "superionic conductors" needs to be complemented by significant advances in the fundamental understanding of ion transport in polymers, where today's theories are not extensive enough to explain many of these properties. Here, extensive use of computer simulations can be increasingly helpful. Moreover, the SPE area has only recently started to use computational tools such as machine learning techniques for true materials design [4]. Molecular dynamics simulations, where atomic transport is targeted at relevant timescales, would be a computational tool of choice to truly capture the mobility mechanisms but need to be coupled also to mesoscopic methods (or better, included in a multiscale model) to capture the decisive microstructures of polymer materials and their physics and chemistry in battery devices. Theories explaining both coupled and decoupled ionic transport in polymeric materials will be decisive for continued exploitation of SPEs in a targeted fashion.

It is also clear from the summary here that the ability for lithium salt dissolution is of fundamental importance for SPE hosts. This has also been one of the strongest arguments for using PEO in the past. However, as seen from the comparatively good performance from many of the alternative host materials, this ability can actually be both a blessing and a curse, since it also leads to excessively strong complexation of the cations. The lower solvation strength of polyesters and polycarbonates renders much better cationic transport than in polyethers – and therefore better transference numbers – because Li^+ is less strongly bound to the polymer. Here, theories better elucidating the interplay between coordination strength, donor number and dielectric properties will be key for the fundamental understanding and design of SPEs. Work in this area is emerging by employment of molecular dynamics simulations [5], but more efforts – also experimental – are clearly necessary for understanding the fundamental physical chemistry of SPE materials.

It can also be noted that, in these host materials, the conductivity maximum due to Li^+-induced chain stiffening is often observed at higher salt concentrations than in PEO, enabling the efficient use of high salt concentrations as a means to increase conductivity. As shown in some of the examples discussed above, the less coupled the ionic motion is to the polymer backbone, the higher the conductivity can become. Here, it seems vital to point out that it is especially the chelating effect of joint ether oxygens in the same polymer chain that seems to have the strongest impact on the limited Li^+ mobility. PEO, or very long oxyethylene side chains that can effectively trap Li^+, might in this context represent a "dead end" for SPE improvement, but if the ether oxygens can be separated by modifying the polymer architecture – including using other types of polymer hosts in combination with the polyethers – progress can be realized, not least regarding improved transference and transport numbers. Polyether segments can thus positively influence both macromolecular mobility and provide decent coordination sites. A rejuvenation can thus be foreseen for polyethers, not least considering their good stability with metallic lithium.

Apart from the polymer, the other crucial component in an SPE is the salt. While this book has not put a very strong emphasis on the salt component, its interplay with the polymer and other battery components is indeed vital for performance. As addressed in Chapter 2, low-lattice-enthalpy salts with plasticizing anions have made a large impact on the field, and further improvements in this area can hopefully appear. Perhaps even more interesting is the design of anions that have a stronger interaction with the polymer host. As mentioned, fluorinated salts that bond to fluorinated backbones in terms of fluorophilic interactions can boost cation transport numbers and diminish concentration polarization. Fluorine-free salts, at the other end of the chemical spectrum, have likely advantages in terms of environmental friendliness and ease of recycling, provided that corrosion problems of such systems are mitigated. Since salt components are also frequently found in the SEI layers of SPEs, there is much tailoring that can be envisioned, and several novel salts are also emerging for SPE applications [6]. Salt anions with affinity for more conventional polymer hosts than fluorinated can also be envisioned – that is, tailoring the salt rather than the polymer for strong anion–polymer interactions.

While we have so far avoided polymer-based composites with ceramic materials in this discussion, this is a category of materials that holds much promise. As also mentioned in Chapter 1, it is not far-fetched to imagine combining the benefits of the high conductivity of ceramics with the better processing and wetting properties of polymers. While delving into these systems is outside the boundaries of this book, it is likely a future prosperous use for SPE materials. It is interesting to note, though, that many of these ceramic–polymer composite electrolytes share properties that are common for SPE composites with ionically inert nanoparticles and that have been well explored since the 1990s. This signals that it is primarily the polymer phase which is responsible for ion transport in such composites, while the ceramic conductors change the polymer matrix and/or provide alternative transport paths for the ionic transport. If this is the major difference with pure SPEs, however, the high conductivity sometimes seen is – somewhat counterintuitive – *not* due to the good ionic transport properties of the ceramic conductor materials, but due to other effects. This area needs considerably more research to elucidate the transport mechanisms, both theoretically and experimentally.

Finally, it is primarily the last 10 years that have seen extensive implementation of polymer electrolytes into battery devices, both in academic research and in industrial settings. The field has now reached the maturity that calls for more extensive standardization of the testing methodology, here described in Chapters 3 and 4, in order to identify pitfalls and make conclusive statements on performances. Such standardized efforts are currently being implemented globally for the general battery area, for example, within the Battery2030+ initiative. It is obvious that too few studies in the past have given good estimations on electrochemical stability, purity of the samples, functionality with electrodes, possible processability, etc. Method improvement is therefore likely to render a significant impact on this research field. Also correlated with the maturity of the technological field, there are factors regarding sustainability and recyclability coming into play, and where SPE-based batteries need to undergo the same kind of environmental analysis and design for end-of-life processing as the more established Li-ion battery chemistries. These are questions of large societal importance, and which have only been rudimentarily analyzed so far. There is plenty more to do for the exploration of polymer-based solid-state batteries.

References

[1]   Zhou W, Wang Z, Pu Y, et al. Double-layer polymer electrolyte for high-voltage all-solid-state rechargeable batteries. Adv Mater. 2019;31:1805574.

[2]   Mankovsky D, Lepage D, Lachal M, Caradant L, Aymé-Perrot D, Dollé M. Water content in solid polymer electrolytes: The lost knowledge. Chem Commun. 2020;56:10167–70.

[3]   Angell CA. Polymer electrolytes – Some principles, cautions, and new practices. Electrochim Acta. 2017;250:368–75.

[4]   Hatakeyama-Sato K, Tezuka T, Umeki M, Oyaizu K. AI-assisted exploration of superionic glass-type Li+ conductors with aromatic structures. J Am Chem Soc. 2020;142:3301–5.

[5]   Wheatle BK, Lynd NA, Ganesan V. Effect of polymer polarity on ion transport: A competition between ion aggregation and polymer segmental dynamics. ACS Macro Lett. 2018;7:1149–54.

[6]   Qiao L, Oteo U, Zhang Y, et al. Trifluoromethyl-free anion for highly stable lithium metal polymer batteries. Energy Storage Mater. 2020;32:225–33.

Index

https://doi.org/10.1515/9781501521140-007